Table of Contents

INTRODUCTION

"A nation that destroys its soil destroys itself. Forests are the lungs of our land, purifying the air and giving fresh strength to our people." – Franklin D. Roosevelt.

"We do not inherit the earth from our ancestors, we borrow it from our children." - Native American Proverb.

"Clean air and water, and a livable climate are inalienable human rights. And solving this crisis is not a question of politics. It is our moral obligation – if admittedly a daunting one." – Leonardo DiCaprio.

Imagine waking up to orange skies as wildfires rage just miles away. Picture coastal towns swallowed by rising tides and entire ecosystems vanishing before our eyes.

The past decade has seen record-breaking temperatures, devastating wildfires, and rising sea levels threatening entire coastal communities. Climate change is no longer a distant threat; it is happening now.

This book unpacks the science behind climate change, explores its impact on daily life, and provides practical solutions. It is an important book for policymakers, students, or anyone who cares about the planet.

This book covers:

- A clear breakdown of what climate change is and why it's happening.
- We're already facing consequences—rising temperatures, extreme storms, and vanishing coastlines.
- How individuals, communities, and governments can take action to slow the crisis and adapt to new realities.

This book is designed for beginners and those with some experience in climate data. Let's begin, shall we?

The Root of the Problem

Wherever you live, you've likely felt the effects of climate change. These include unpredictable droughts, destructive floods, extreme storms, or record-breaking heatwaves. These events, once rare, are now alarmingly frequent.

Oceans are also changing. Warming temperatures and increasing acidification are harming marine life, including coral reefs. When these signs appear together, they strongly indicate climate change (see Chapter 1 for more details).

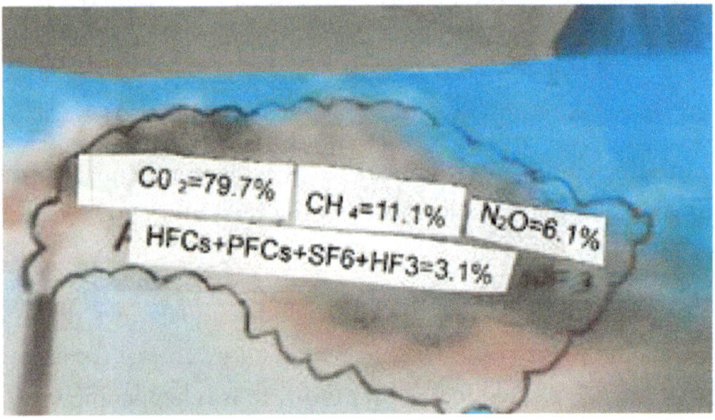

However, scientists have long debated Earth's natural climate cycles, but overwhelming evidence now shows that human activities, especially the burning of fossil fuels, are driving unprecedented changes.

Hence, the main responsibility falls on the shoulders of humans or the humans burning fossil fuels. These fossil fuels in the form of coal, oil and natural gas are formed naturally in the earth's crust from the remains of prehistoric organisms. This geological process takes millions of years (see Chapter 3).

Burning of these fossil fuels accumulates greenhouse gases, such as carbon dioxide (CO_2), methane (CH_4), nitrous oxide (N_2O), ozone (O_3), and fluorinated gases in the Earth's atmosphere (see Chapter 4).

Think of Earth as a giant greenhouse. The sun's rays enter, warming the surface, but instead of escaping back into space, heat gets trapped by a layer of gases like carbon dioxide and methane. These gases act like an invisible blanket, keeping the planet warm, sometimes too warm.

Since the Industrial Revolution (circa 1760–1840), human activities have released increasing amounts of greenhouse gases. Emissions surged in the 20th century, rising by 70% between 1970 and 2004, with CO_2 alone increasing by 80% during that period.

Despite these challenges, humanity has begun to act:

- Reducing reliance on fossil fuels.
- Expanding clean energy sources like solar, wind, and hydropower.
- Collaborating globally through initiatives like the United Nations' climate agreements (see Chapter 5).

This crisis demands action from all of us, whether through personal choices, advocacy, or innovation. The question is: what role will you play?

CHAPTER 1
CLIMATE CHANGE

Understanding the root causes of climate change is the first step toward finding solutions. Let's begin by exploring what climate change is and why it matters.

What Is Climate Change?

It refers to long-term shifts in weather patterns and temperature due to human activities, distinguished from natural causes.

One of the most obvious examples of these human activities is burning fossil fuels such as coal, oil, and natural gas. Since the industrialization of the 1800s, burning fossil fuels has resulted in heat-trapping gases in the atmosphere, also called "greenhouse gases."

The result? The air we breathe is changing. Greenhouse gases are also increasing rapidly:

- Carbon dioxide (CO_2) is the largest share (79.4%).
- Methane (CH_4) follows at 11.5%, trapping heat more effectively than CO_2.
- Nitrous oxide (N_2O) at 6.2% and fluorinated gases (3%) are small in quantity but have a powerful warming effect.

Though present in smaller amounts, methane and fluorinated gases trap far more heat than CO_2, accelerating global warming.

These gases are so named because they act like the glass walls of a greenhouse, trapping the sun's heat.

But how do we know this isn't just natural change? The evidence is overwhelming:

1. According to the Intergovernmental Panel on Climate Change (IPCC), the influence of human activity on global warming is an established fact.

2. The earth's average surface temperature has risen by about two degrees Fahrenheit (one degree Celsius) since the late 19th century – a change driven largely by increased carbon dioxide emissions into the atmosphere.

3. Most of the warming occurred in the past 40 years, with the seven most recent years being the warmest.

4. The years 2016 and 2020 are tied as the warmest years on record. Scientists at NASA's Goddard Institute for Space Studies (GISS) found that the global average temperature for 2020 was 1.84 degrees Fahrenheit (1.02 degrees Celsius) above the 1951-1980 baseline. The difference falls within the margin of error, making the two years the hottest on record.

5. The oceans have absorbed much of the rise in heat, about 90%, with the result that the top 100 meters (328 feet) have warmed by an extra 0.67 degrees Fahrenheit (0.38 degrees Celsius) since 1969.

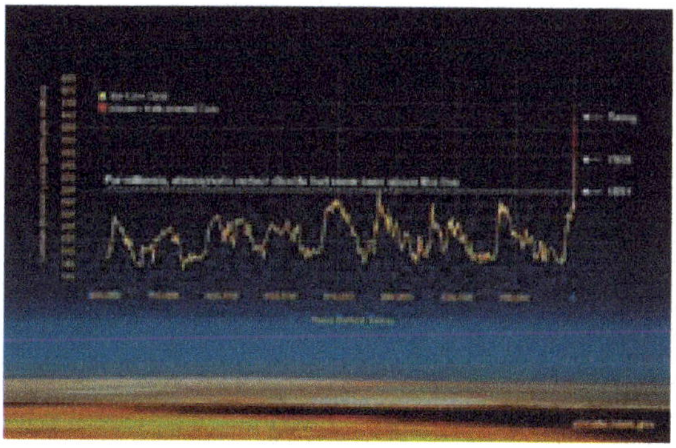

Graph credit: climate.nasa.gov/ public domain

6. The oceans are also rising. They rose by about 20 centimeters (8 inches) in the last 100 years. However, the rise has nearly doubled over the previous 20 years compared to the rise in the last 10 years. Also, the rise is accelerating in each subsequent year.

So, climate change is real, and so are its effects:

1. Warmer temperatures over time are changing weather patterns and disrupting the usual balance of nature. This poses risks to humans and animals due to habitat loss by wildfires that start more easily during hot weather.

2. Scorching heatwaves aren't just uncomfortable—they're deadly. Rising temperatures are causing more heatstroke cases and making outdoor work hazardous.

3. More severe storms: changes in temperature causes changes in rainfall, resulting in more frequent and intense storms, which in turn causes flooding and landslides. Homes and communities are destroyed costing billions of dollars.

4. More severe drought: water is a scarcer commodity in relatively hotter regions. Many communities have reduced or no water supply. Vegetative cover can disappear, and the land is subject to wind-related erosion resulting in desert-like conditions such as dust storms. Arable areas for agriculture and grazing areas for cattle can be drastically reduced.

5. Oceans are absorbing excess heat, melting ice sheets at an alarming rate. Coastal cities are already losing ground to the rising tide, with some islands facing total submersion.

6. Loss of species: climate change poses risks for the survival of species on land and in the ocean. These risks increase as temperatures climb. Forest fires, extreme weather, and invasive pests and diseases are among many threats. Some species will be able to relocate and survive, but others will not be able to.

7. Not enough food: changes in climate and increases in extreme weather events are among the reasons behind a global rise in hunger and poor nutrition. Fisheries, crops, and livestock may be destroyed or become less productive. Heat stress can diminish water and grasslands for grazing.

8. More health risks: changing weather patterns are expanding diseases such as malaria. Extreme events increase disease and death and make it difficult for health care systems to keep up. Other risks to health include increased hunger and poor nutrition in places where people cannot grow or find sufficient food.

9. Poverty and displacement: climate change increases the factors that put and keep people in poverty. Floods may sweep away urban slums, destroying homes and means of livelihood. Heat can make it difficult to work outdoor jobs. Weather-related disasters

displace 23 million people a year, leaving many more vulnerable to poverty.

10. Ocean acidification has been increasing since the beginning of the Industrial Revolution (1760-1840), the acidity of the surface ocean waters has increased by about 30%. This increase is due to human activities emitting more carbon dioxide into the atmosphere and hence more being absorbed into the ocean. The ocean has absorbed between 20-30% of total carbon dioxide emission in recent decades (7.2 to 10.8 billion metric tons per year).
11. Arctic Sea ice has been declining in both areal extent and thickness in the last several decades.
12. Snow cover is decreasing. Satellite observations reveal that the snow cover in the Northern Hemisphere has decreased over the past five decades and the snow is melting earlier.
13. Glaciers are retreating. Earth's glaciers are retreating almost everywhere around the world including the Alps, Himalayas, Andes, Rockies, Alaska, and Africa. What is a glacier? A glacier is a large, perennial accumulation of crystalline ice, snow, rock, sediment, and often liquid water that originates on land and moves downslope under its own weight by gravity.
14. Extreme events are increasing in frequency: The number of record high temperature events in the United States has been increasing while the number of record low temperature events has been decreasing since 1950. The United States has also witnessed increasing numbers of intense rainfall events. In a likely scenario, humans with their flora and fauna could be moved to Mars, as scientists have long speculated as Mars is quite like Earth in many respects.

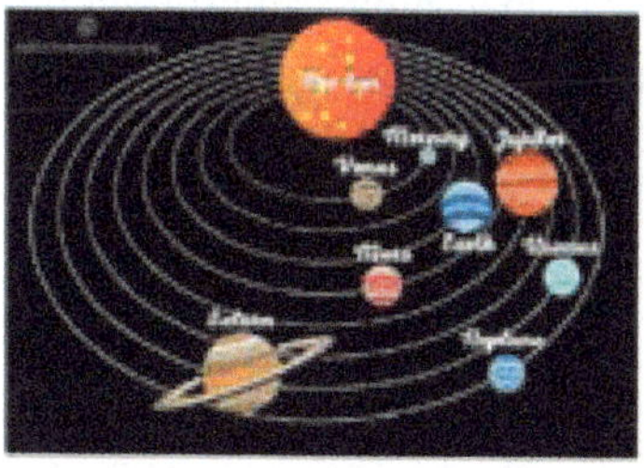

Image from Public Domain

Top 10 Countries Vulnerable To Climate Change

These countries face severe risks from climate change due to a combination of extreme weather events, environmental degradation, economic instability, and limited adaptive capacity:

1. Democratic Republic of Congo

Increasing deforestation, erratic rainfall, and extreme weather threaten livelihoods and biodiversity in one of the world's largest rainforest regions.

2. Central African Republic (CAR)

Already grappling with conflict and food insecurity, the country faces rising temperatures, prolonged droughts, and unpredictable rainfall patterns that worsen agricultural challenges.

3. Nigeria

Flooding, desertification, and rising sea levels pose major threats, especially in low-lying coastal regions and the drought-prone northern areas.

4. Chad

Desertification and declining water resources, particularly in the shrinking Lake Chad Basin, put millions at risk of displacement and food shortages.

5. Ethiopia

Severe droughts and erratic rainfall patterns threaten agricultural production, exacerbating hunger and economic instability.

6. Bangladesh

One of the most flood-prone countries in the world, Bangladesh faces rising sea levels, cyclones, and extreme weather that threaten millions living in low-lying coastal areas.

7. Syria

Prolonged droughts and water shortages have worsened socio-political instability, contributing to food crises and mass displacement.

8. Afghanistan

Water scarcity, desertification, and increasing temperatures make agricultural production highly unpredictable, intensifying food insecurity and displacement.

9. Somalia

Recurring droughts, rising temperatures, and erratic rainfall patterns contribute to severe famine risks, forced migration, and conflict over scarce resources.

10. South Sudan

Floods and droughts are becoming more frequent, threatening food production and intensifying humanitarian crises in a country already dealing with instability.

All these countries are unprepared for events like flooding, drought, food scarcity, and disease outbreak.

Top 5 CO_2-Emitting Countries

The following five countries are the largest contributors to global carbon dioxide (CO_2) emissions, primarily due to their reliance on fossil fuels for energy and industrial production.

1. China

The world's largest CO_2 emitter, China released 11,397 million metric tons of carbon dioxide in 2022. Its emissions are primarily driven by coal, which accounts for 58% of the country's energy production. Coal-fired power plants and industrial operations release vast amounts of CO_2 into the atmosphere. Additionally, China is a major oil importer, with transportation contributing significantly to its carbon footprint. However, to curb pollution in major cities, China has been investing in nuclear energy, renewables, and natural gas to reduce its dependence on fossil fuels.

2. United States

As the second-largest emitter, the U.S. released 5,057 million metric tons of CO_2 in 2022. The primary sources of emissions include transportation, electricity production, and heavy industry. While the

U.S. has made efforts to shift away from coal, it remains a major crude oil producer, fueling its energy and transportation sectors. Passenger vehicles, trucks, ships, trains, and planes consume large amounts of petroleum, making transportation a leading contributor to emissions. The chemical industry also adds significantly to the U.S. carbon footprint, as fossil fuels are used in manufacturing processes that release CO_2.

3. India

Ranking third, India emitted 2,830 million metric tons of CO_2 in 2022. Coal is its primary energy source, supplying 44% of the nation's power. Petroleum and other liquids contribute 24%, while natural gas accounts for only 6% of India's energy consumption. However, India has set a goal to increase natural gas usage to 15% by 2030 as part of its effort to transition to cleaner fuels and reduce air pollution.

4. Russia

As the fourth-largest CO_2 emitter, Russia produced 1,652 million metric tons in 2022. The country has vast natural gas reserves, making it a dominant fuel for power generation and industry. Coal remains widely used in Russia's chemical and manufacturing sectors, further adding to emissions. Russia has historically been a major natural gas supplier to European nations, but due to geopolitical tensions, particularly following its invasion of Ukraine, European countries have begun reducing reliance on Russian gas, while Russia has also cut back on exports.

5. Japan

Emitting 1,054 million metric tons of CO_2 in 2021, Japan is the fifth-largest carbon emitter. The country's energy mix shifted dramatically after the 2011 Fukushima nuclear disaster, leading to increased reliance on fossil fuels. Oil remains Japan's largest energy source at 38%, followed by coal (25%) and nuclear power (23%). Japan's energy strategy aims to increase nuclear power capacity by 2030 to lower its dependence on imported fossil fuels and reduce overall emissions.

CHAPTER 2
GLOBAL WARMING

Now that we've covered climate change let us talk about the phenomenon of global warming.

Even though global warming and climate change are used interchangeably, they are not the same.

- Global warming refers to the steady rise in Earth's temperature, mainly due to greenhouse gas emissions.
- Climate change is the broader shift in weather patterns—stronger storms, shifting rainfall, and extreme droughts—happening over decades or centuries.

The Paris Agreement aims to keep global temperature rise well below 2 degrees Celsius above pre-industrial levels (1760–1840) by the end of this century. Additionally, countries committed to making further efforts to limit warming to 1.5 degrees Celsius, recognizing that even small increases in temperature could lead to severe environmental and socio-economic consequences. More on this in Chapter 7.

For now, remember that crossing 1.5°C is not just a milestone—it's a turning point:

- More extreme storms and wildfires will devastate communities.
- Sea levels will rise, displacing millions.
- Coral reefs and marine ecosystems could collapse.
- Heatwaves and droughts will make many regions unlivable.

If temperatures exceed 31°C (88°F), people cannot perform physical labor over long periods without putting their health at serious risk.

And If temperatures surpass 35°C (95°F), a healthy human can die within a few hours without access to water or shelter.

How Hot Will the Earth Be in 2030?

Most climate models from the United Nations Intergovernmental Panel on Climate Change (IPCC) project that global temperatures will breach the 1.5°C threshold above pre-industrial levels sometime during the 2030s.

Can Global Warming Cause More Frequent and Intense Wildfires?

While global warming does not directly cause wildfires, research shows a strong correlation between rising temperatures and the frequency and intensity of wildfires. Hot, dry conditions create the perfect environment for fires to spread, making them more destructive. Though some fires occur naturally due to lightning strikes, many are started by human activity, either accidentally or intentionally.

At the heart of global warming lies the burning of fossil fuels. Let's examine their origins, their role in climate change, and the challenges of moving away from them.

CHAPTER 3
FOSSIL FUELS

Fossil fuels—coal, oil, and natural gas—are the buried remains of ancient life, compressed over millions of years. But burning them releases carbon that has been locked away for ages, disturbing Earth's delicate balance.

Among these, coal and petroleum have been formed from the remains of dead plants and animals subjected to various biological and geological processes for millions of years. Coal is a natural resource that is obtained from the remains of trees and other plants that lived millions of years ago, and petroleum is obtained from the remains of dead animals that also lived millions of years ago.

In contrast, natural gas is a lighter, gaseous fossil fuel. Composed mainly of methane (CH_4), it is odorless in its natural state and is often found alongside oil and coal deposits. Though considered a "cleaner" fossil fuel due to lower carbon dioxide emissions when burned, natural gas still contributes significantly to climate change, especially when methane leaks into the atmosphere, as it is far more potent at trapping heat than CO_2.

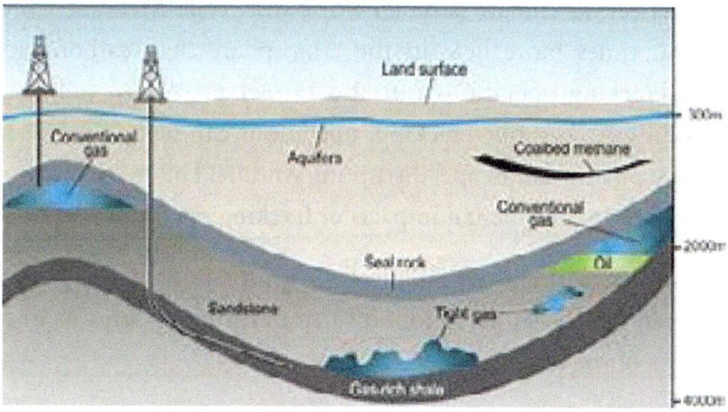

How Is Natural Gas Formed?

As the organic matter moves deeper into Earth's crust, it encounters higher and higher temperatures. The combination of compression and high temperature causes the carbon bonds in the organic matter to break down. This molecular breakdown produces thermogenic methane – natural gas.

All About Fracking

Fracking, or hydraulic fracturing, is a process where rock deep underground is shattered to extract oil and gas. But this method has a hidden cost—it releases large amounts of methane, a greenhouse gas 80 times more potent than CO_2.

Its effect on climate is much worse than coal. Methane gas traps about 86 times more heat in the atmosphere than carbon dioxide (CO_2). Fracking operations can also extract petroleum (oil) instead of natural gas. Without Fracking, operations can be an eyesore in an otherwise pristine landscape, disrupting wildlife. Following are some of the negative environmental impacts of fracking operations.

1. Fracking a single well typically requires millions of gallons of water, which depletes groundwater and dries up streams in some parts of the country.
2. Fracking creates massive volumes of toxic and sometimes radioactive wastes
3. Public Domain Images

4. (deep drilling may encounter radium). Conventional treatment plants are not equipped to treat fracking wastewater. Getting rid of it by injecting it underground could cause an earthquake and jeopardize drinking water.

5. Corporations don't have to disclose what chemicals there are in their fracking fluid. The toxic chemicals and mixtures are called "trade secrets," but among those known to be used, they may cause cancer, neurological problems, and wide-ranging health and environmental effects.

6. Fracking is linked to birth defects. A landmark study conducted by researchers at the Colorado School of Public Health and Brown University found that babies born to mothers who lived in areas dense with gas wells were 30 percent more likely to have congenital birth defects.

7. Fracking causes thousands of accidents, leaks, and spills every year.

8. Fracking corporations are exempt from critical statutes in the Safe Drinking Water Act, the Clean Water Act, the Clean Air Act, and laws regarding the disposal of toxic waste.

9. Methane leaks from drilling and fracking have caused houses and water wells (not just oil and gas wells) to explode, causing deaths, injuries, and loss of property.

10. Trains carrying shale oil from drilling and fracking have exploded, causing widespread destruction and death. Also, tanks holding oil and gas industry wastes have been struck by lightning and exploded, spilling contaminants onto surrounding lands and burning for days.

11. Fracking is expanding at an alarming rate. Fracking corporations are banking on being able to drill and crack for as much oil and gas as possible, including in national parks and densely populated areas of the country.

12. Fracking locks in utterly disruptive climate change. Fracking wells release large amounts of methane gas, exacerbating global warming.

(**Source**: Food and Water Watch, a nonprofit organization that advocates safe food, clean water, and a livable climate).

Several countries have taken legal steps to ban or restrict fracking due to environmental concerns. In some cases, outright bans exist, while in others, moratoriums prevent new fracking projects from moving forward.

Countries With a Complete Fracking Ban

- France (2011) – The first country to ban fracking, citing environmental risks. In 2017, France extended its commitment by passing a law to ban all oil production by 2040.
- Germany (2017) – Banned unconventional fracking to protect the environment.
- Ireland – Prohibited new licenses for oil and gas exploration.
- Spain (2021) – Approved a climate law that aims to phase out fossil fuel production by 2042.
- Denmark – Agreed to stop issuing new oil and gas exploration licenses in the North Sea, with a full phase-out planned by 2050.
- Belize (2018) – Passed legislation to end oil exploration and extraction in its territorial waters.

Countries With Moratoriums or Restrictions

- Tunisia – Implemented restrictions on fracking, though not an outright ban.
- Quebec, Canada – Enforced a de facto moratorium, meaning fracking is effectively halted despite the absence of a formal legal ban.

Landfills' Impact on Climate

Landfills are a primary method of waste disposal worldwide, but their impact on the environment is growing at an alarming rate. The sheer volume of waste continues to rise, posing significant challenges for sustainability. According to the World Bank, global waste production reached 2.0 billion tons in 2016. If current trends continue, this number could soar to 3.4 billion tons by 2050—a nearly 70% increase in just three decades.

Reports from Sustainability for All and the International Solid Waste Association (ISWA) highlight the environmental consequences of landfills, particularly their role in contributing to climate change through methane emissions and long-term pollution.

Landfills are one of the causes of climate change as these release biogas into the atmosphere. Biogas is a mixture of methane gas (CH_4) and carbon dioxide (CO_2) – two of the gases that cause climate change and global warming. According to a report by ISWA (International Solid Waste Association) called "A Roadmap for Closing Waste Dumpsites," determined that landfill sites currently contain approximately 40% of the world's waste, and the 50 largest landfill sites impact the daily lives of 64 million people - equivalent to the entire population of France.

Helpfully, many communities have separated recyclable waste and treated it in recycling plants. Thus, much of the waste we generate takes on a second life, though, unfortunately, this is not the standard practice. Most waste ends up in landfills.

How do landfill sites work? A landfill site is shaped (below or above existing grade) and covered with an artificial water-resistant coating to prevent the landfill from contaminating the soil. The site should be isolated (away from residential, commercial, and industrial areas). The waste is crushed and compacted to reduce the fill volume and transported to the landfill site, where it is dumped and buried under soil layers and left to decompose. However, the waste takes many years to decompose as there is little oxygen and moisture.

Why are landfill sites so harmful to the environment? Landfill sites are one of the causes of climate change. They are partially responsible for global warming by generating and releasing biogas into the atmosphere. Biogas is a mixture of methane gas (CH_4) and carbon dioxide (CO_2), two of the gases that cause climate change and an increase in the planet's temperature. According to the ISWA report, if the current situation continues and we do not act, landfill sites will account for 10% of greenhouse gas emissions by 2025.

Landfill sites can cause fires or explosions. Sometimes, methane produced by landfill waste can cause explosions and fires. This downside is more common than meets the eye because the fires that occur are not regular fires with flames but fires that occur inside the landfill. Dioxin emissions from these spontaneous uncontrolled fires are also very harmful to the environment, not to mention the damaging effects they have on aquifers, whose waterproof membranes are affected by the fire.

Landfill sites can contaminate soil or water. They are often responsible for contamination of soil and groundwater, as the contaminating materials (such as heavy materials like lead and mercury) that the stored waste can contain can spread to the soil and water near plants. Moreover, although it is not very common for waterproofing membranes to rupture, they have a devastating effect on the environment when they do.

Landfills can affect the migratory behavior of birds such as cranes, Canadian geese, hummingbirds, and Arctic Terns. Over 300 million birds start their annual fall migration to the south for the winter. These birds are attracted by the abundant food resources of the landfills, which unfortunately provide a deadly diet of plastic, aluminum, gypsum, and other materials in the waste. Moreover, these birds have altered their migratory activity, nesting near landfills.

Landfills reduce the value of the surrounding areas. The foul smells that emanate from the landfills and those that cannot be controlled inevitably make people leave these areas.

Accidents sometimes occur in landfill areas. In March 2017, the Addis Ababa landfill site in Ethiopia collapsed, causing an estimated 113 deaths. Only a month later, the Meethotamulla landfill site in Sri Lanka suffered a landslide, leaving more than 30 people dead, dozens of people missing, and more than 140 houses destroyed. In February 2020, two workers were killed when the Zaldivar landfill site in Spain collapsed. Rain, spontaneous combustion, or excessive accumulation sometimes turn landfill sites into unstable terrain where landslides or collapses are dangerous for plant workers and nearby areas.

The Impact of Plastics

Plastic pollution is a climate crisis in disguise:

- 99% of plastic comes from fossil fuels.
- Its production, use, and disposal release greenhouse gases at every stage.
- If plastic use continues unchecked, emissions could reach 56 gigatons by 2050—10-13% of the remaining global carbon budget.

The Plastic Production Process

The manufacturing of plastic follows a multi-step process, beginning with raw fossil fuels and ending with the production of plastic products.

1. Extraction – Crude oil and natural gas are drilled from the ground and transported to a refinery.
2. Refining – The raw materials undergo processing to separate useful components.
3. Polymerization – A chemical reaction converts monomers (small molecules) into polymers (long-chain molecules), forming different types of plastic:
    - Polyethylene – Used in plastic bottles and packaging.
    - Polyvinyl chloride (PVC) – Used for plastic pipes.
    - Polypropylene – Used in containers, textiles, and automotive parts.
4. Nurdle Formation – The resulting plastic is shaped into tiny pellets known as nurdles, which serve as the building blocks for plastic products such as bags, bottles, and packaging materials.

The plastic pollution crisis is not just an environmental issue—it is a growing threat to the Earth's climate. At current levels, greenhouse gas emissions from the plastic lifecycle endanger global efforts to keep temperature rise below 1.5°C. With the petrochemical and plastic industries planning massive expansions in production, this crisis is set to escalate.

What Will Happen If Plastic Production Continues to Grow?

If plastic production and use continue as planned:

- By 2030, global greenhouse gas emissions from plastics could reach 1.34 gigatons per year—equivalent to the emissions from over 295 new 500-megawatt coal-fired power plants.
- By 2050, cumulative plastic-related emissions could surpass 56 gigatons, accounting for 10-13% of the remaining global carbon budget.

How Does the Plastic Lifecycle Contribute to Climate Change?

Nearly every piece of plastic begins as a fossil fuel, and greenhouse gases are emitted at every stage of its lifecycle:

1. Fossil fuel extraction and transport – Oil and gas drilling release methane, carbon dioxide, and other pollutants.
2. Plastic refining and manufacturing – Energy-intensive processes generate significant emissions, particularly from cracking alkanes into plastic resins.
3. Plastic waste management – Landfilling, recycling, and incineration all contribute to emissions, with incineration being the most carbon-intensive.
4. Environmental impact – Once released into the environment, plastic continues to affect the climate by breaking down and releasing methane and other gases.

What Are the CIEL Report's Recommendations?

The Center for International Environmental Law (CIEL) outlines key recommendations for policymakers, governments, nonprofits, and other stakeholders:

- Immediately reduce plastic production and use, particularly single-use plastics.
- Halt the expansion of petrochemical and plastic production to keep fossil fuels in the ground.
- Promote zero-waste communities to reduce dependence on landfill and incineration.
- Enforce extended producer responsibility as part of a circular economy model.
- Set ambitious emission reduction targets across all sectors, including plastic production.

Why Are Climate Targets Hard to Meet?

Greenhouse gas emissions from plastics are making it increasingly difficult to meet global climate goals:

- By 2019, plastic production and incineration added over 850 million metric tons of CO_2 to the atmosphere—equivalent to the emissions of 189 coal power plants.
- At current rates, these emissions jeopardize the world's ability to meet carbon reduction targets.

How Much Do Extraction and Transport Contribute to Emissions?

The extraction and transport of fossil fuels for plastic production generate substantial emissions. Sources include:

- Methane leakage and flaring during oil and gas drilling.
- Energy consumption in extraction and refining.
- Land disturbance from clearing forests and fields for drilling operations.
- In 2015, U.S. emissions from plastic-related fossil fuel extraction totaled 9.5–10.5 million metric tons of CO_2 equivalent per year. Globally, 108 million metric tons of CO_2 emissions were linked to plastic production.

What Are the Emissions from Plastic Refining and Manufacturing?

Plastic refining is one of the most greenhouse-gas-intensive industries and is rapidly growing.

- In 2015, U.S. ethylene plants produced 17.5 million metric tons of CO_2, equal to the emissions of 3.8 million passenger vehicles.
- Globally, emissions from ethylene production ranged between 184.3 – 213.0 million metric tons of CO_2—comparable to 45 million cars driven for a year.
- New facilities, such as Shell's ethane cracker in Pennsylvania and ExxonMobil's Baytown plant in Texas, will add millions of tons of CO_2 emissions annually.

How Is Plastic Waste Managed?

Plastic waste is typically:

- Landfilled – Produces the least emissions but poses long-term environmental risks.
- Recycled – Moderately reduces emissions but does not eliminate the need for virgin plastic.
- Incinerated – The most carbon-intensive method, responsible for rising emissions.

By 2015, emissions from plastic incineration in the U.S. were estimated at 5.9 million metric tons of CO_2. Global emissions from burning plastic packaging alone reached 16 million metric tons. This figure does not include open burning, which releases even more carbon.

What Happens When Plastic Ends Up in the Environment?

Unmanaged plastic continues to contribute to climate change as it degrades.

- A first-of-its-kind study found that plastic at the ocean's surface continuously releases methane and other greenhouse gases.
- Emissions increase as plastic breaks down further in landfills, rivers, and coastlines.
- Microplastic pollution in the ocean disrupts the carbon cycle, interfering with the ability of phytoplankton to absorb CO_2.
- Zooplankton, which helps transport carbon to deep ocean layers, are harmed by microplastic contamination, reducing their ability to sequester carbon.

Can We Still Limit Global Temperature Rise to 1.5°C?

The expansion of plastic production threatens to push global temperatures beyond the 1.5°C target, as mentioned in the previous chapter.

- If plastic production, disposal, and incineration continue at their current trajectory:

  - By 2030, emissions could reach 1.34 gigatons per year, equivalent to 295 new coal-fired power plants.
  - By 2050, plastic-related emissions could reach 2.8 gigatons per year, the same as 615 coal plants.
- Cumulative emissions by 2050 could exceed 56 gigatons of CO_2, accounting for 10-13% of the remaining global carbon budget.
- If this trend continues, by 2100, plastic-related emissions could reach 260 gigatons of CO_2, over half of the total remaining carbon budget.

How Can We Address the Plastic Pollution Crisis?

The CIEL report identifies urgent actions needed to mitigate plastic's climate impact:

- Ban single-use plastics to reduce unnecessary waste.
- Stop new oil, gas, and petrochemical infrastructure that supports plastic production.
- Encourage zero-waste communities to minimize reliance on landfills and incineration.
- Implement extended producer responsibility policies to hold companies accountable for plastic waste.
- Set and enforce strict greenhouse gas reduction targets across all industries.

While some solutions, such as bio-based plastics, are being explored, they often fail to address plastic's full climate impact and may create new environmental problems. Only a systemic reduction in plastic production and fossil fuel use can significantly curb emissions and limit climate change.

CHAPTER 4
GREENHOUSE GASES

Greenhouse gases are the invisible drivers of global warming. They trap heat in Earth's atmosphere, creating what is known as the "greenhouse effect."

Without these gases, our planet would be too cold to support life. But in excess, they disrupt Earth's energy balance, leading to rising temperatures, melting ice, stronger storms, and rising seas.

The major greenhouse gases include:

- Carbon dioxide (CO_2)
- Methane (CH_4)
- Nitrous oxide (N_2O)
- Perfluorinated compounds (PFCs)
- Hydrofluorinated compounds (HFCs)
- Sulfur hexafluoride (SF_6)
- Nitrogen trifluoride (NF_3)

Each of these gases varies in its ability to trap heat. Some, like carbon dioxide, are released in vast quantities but have a lower warming potential. Others, like methane and fluorinated gases, are released in smaller amounts but are far more potent.

What Are Their Relative Percentages?

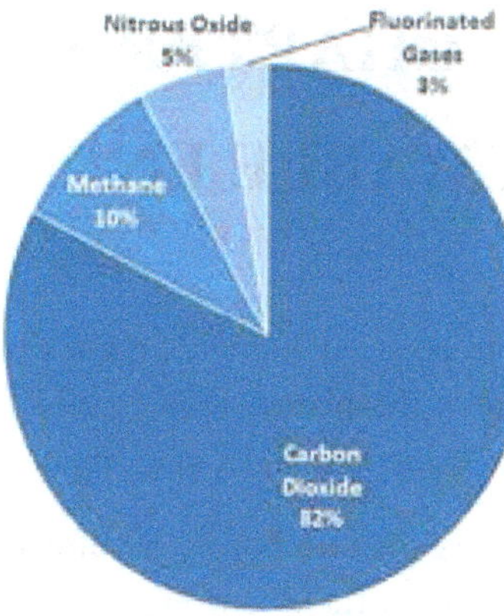

In terms of global emissions, the approximate contribution of each gas to the greenhouse effect is as follows:

- Carbon dioxide (CO_2): ~76%
- Methane (CH_4): ~16%
- Nitrous oxide (N_2O): ~6%
- Fluorinated gases (PFCs, HFCs, SF_6, NF_3): ~2%

Carbon dioxide remains the most significant contributor due to its abundance, primarily from burning fossil fuels. Methane, largely released from agriculture, landfills, and fossil fuel extraction, traps much more heat than CO_2 over a shorter period.

Nitrous oxide mainly comes from agricultural fertilizers and industrial activities. Though released in much smaller amounts, fluorinated gases have global warming potentials thousands of times greater than carbon dioxide and remain in the atmosphere for centuries.

Each gas plays a role, but they are driving the climate crisis forward. In the next chapter, we'll explore how these gases interact with natural and human systems and how carbon footprints reflect their impact on our daily lives.

CHAPTER 5
RENEWABLE ENERGIES

Fossil fuels have powered human progress, but they have also caused serious harm to the planet. Rising emissions, pollution, and warming temperatures now demand a new approach.

Renewable energy comes from natural sources like sunlight, wind, water, and organic matter. These sources do not run out and do not release harmful gases when used.

This chapter introduces the main types of renewable energy and explains how each one can help reduce our dependence on fossil fuels.

Solar Energy

Why is solar energy considered renewable?

Solar panels generate clean, renewable electricity on sunny days, and this energy can be stored in batteries for use during cloudy or rainy weather.

Are solar panels vulnerable to damage?

The answer is yes:

1. Gusty wind can lift solar panels off their restraints and blow them away. So, it is important to secure these on the ground.
2. Wind-driven debris can damage solar panels.
3. Lightning strikes can damage the modules, cables, and electrical equipment.
4. Micro-cracking or micro-fracture can occur in solar panels subject to strong wind force. The silicon used is very thin, and when it expands and contracts or is damaged by wind or falling debris, it can crack, making the panel less efficient in absorbing and storing energy.

5. Rust and water may also travel into the cracks, further harming the effectiveness of the panels.

Source: Holloway Insurance Services. info@hollowayinsuranceco.uk

What is solar energy, and how can it be used?

Solar energy is the sun's light and heat. It can be used in several different ways:

1. Photovoltaic cells, which convert sunlight into electricity.
2. Solar thermal technology, where heat from the sun is used to make hot water or steam.
3. Passive solar heating can be as simple as letting the sunshine through the windows to heat the inside of a building.

Source: archive.epa.gov/public domain.

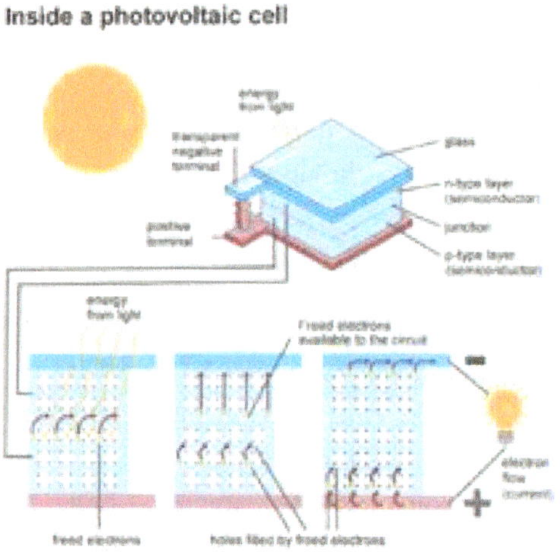

Source: Archive EPA.gov – Public Domain

What is the areal extent of these solar farms?

A solar farm (photovoltaic power station) is a large-scale energy system that converts sunlight into electricity. It has many solar panels (aka photovoltaic or PV modules) connected to form an array. These arrays can cover hundreds of acres of land

What needs to be done to connect the electricity generated by a solar farm to a power grid?

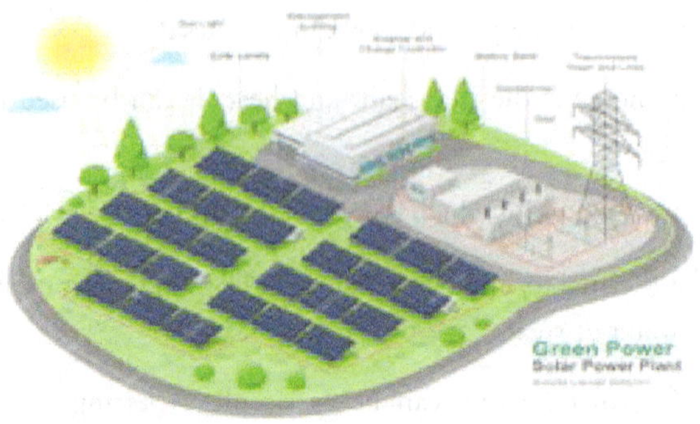

Picture Credit: DOE – Public Domain

Here's what's involved:

1. **Inverters:** These devices convert direct current (DC) produced by the solar panels into alternating current (AC) that the grid can use. Inverters come in different types and sizes to adjust the voltage and frequency of the electricity to match the requirements of the power grid transformers. Once DC is converted to AC, its voltage needs to be increased or decreased depending on the needs of the power grid.

2. **Transmission Lines**: The transferred AC electricity is then fed into the transmission lines, which are large cables designed to transport electricity over long distances. These lines are usually made of aluminum or copper and are supported by tall transmission towers. The electricity travels through these lines at high voltages to reduce energy loss during transmission.

3. **Substations**: Along the transmission lines, substations distribute the electricity at locations as needed. These substations contain additional transformers that can further adjust the voltage of the electricity and route it safely and efficiently to different distribution lines.

4. **Distribution Lines:** From the substations, the electricity is then sent through smaller distribution lines that branch out to

different communities and households. These lines are usually underground or mounted on poles, delivering the electricity directly to consumers' homes and businesses.

While solar panels work best with abundant sunlight, but wind turbines offer another renewable option that draws power from the air.

Wind Energy

What are the components of wind turbines needed to produce electricity?

There are seven major parts and many minor parts. The seven major parts are the tower, the structure's foundation, the wind, the curved blades of the wind turbine, the nacelle, the gearbox, and the generator.

What is the function of the tower?

The tower comprises tubular steel that usually comes in three sections and is assembled on-site. The taller the tower, the better, as wind speeds increase with height. Thus, taller towers enable turbines to capture more energy and generate more electricity.

What are the design considerations for the structure's foundation?

The foundation is a large block of reinforced concrete constructed underground that supports the dead weight of the tower and the turbine and the forces acting on these.

What should be gentle or moderate wind speeds?

Optimal wind speeds range between 12 to 25 kilometers per hour (8 to 16 miles per hour). Stronger winds above 88 kilometers per hour (55 mph) can damage the turbine.

What is the purpose of the turbine's curved blades?

This is similar to the airfoil in an airplane. As can be seen in a cross-section, whereas the upper surface of the blade has a pronounced curvature, the lower surface is more or less horizontal. This creates a longer path for the airstream on the upper surface compared to that on the lower surface. This translates to a relatively high pressure on the lower surface than on the upper surface. In other words, lift L.

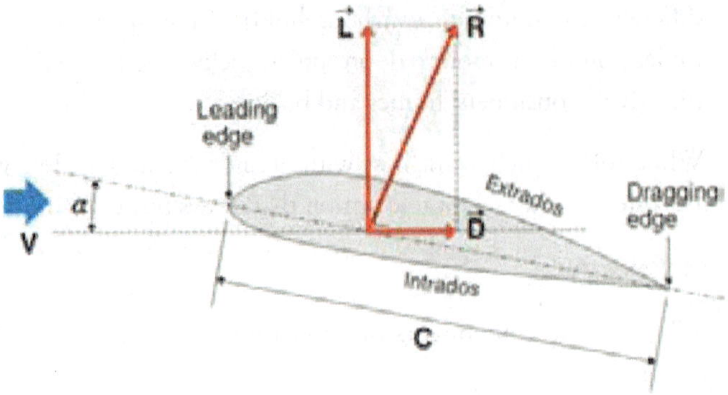

Source: Public Domain

Where wind and sun are intermittent, water provides a more consistent source of renewable energy.

Hydroelectric Energy

Available worldwide, it is the world's oldest form of renewable energy, older than solar, wind, tide, or wave renewable energies.

The oldest known hydroelectric power plant started in 1818 on the Crag side in Northumberland, England.

In the United States (U.S.), what percent of renewable hydroelectric power is supplied by hydroelectricity?

This is about 28.7 percent of total U.S. renewable electricity generation, which is about 6.2 percent of total U.S. electricity generation. (**Source**: U.S. Department of Energy).

What are the common types of hydropower facilities?

There are three types of hydropower facilities:

1. Impoundment
2. Diversion
3. Pumped Storage

The most common hydropower facility is the impoundment type, which creates a reservoir by damming a river. This stored water can be released to generate electricity when needed. For example, the world-famous Hoover Dam built across the Colorado River in Nevada (during 1931-36) created a reservoir known as Lake Mead with a storage capacity of 35.65 billion cubic meters or 28.9 million acre-ft of water. The surface area of Lake Mead is about 642 square kilometers or 248 square miles.

How does an impoundment facility produce electricity?

Sluice gates built in the dam can be opened to let water in the penstocks that rotate a turbine, which in turn rotates magnets inside wires (generator) that create electricity. The electricity is sent to a transformer, which in turn is connected to power transmission cables. The following image adapted from the Hoover Dam cross-sectional view

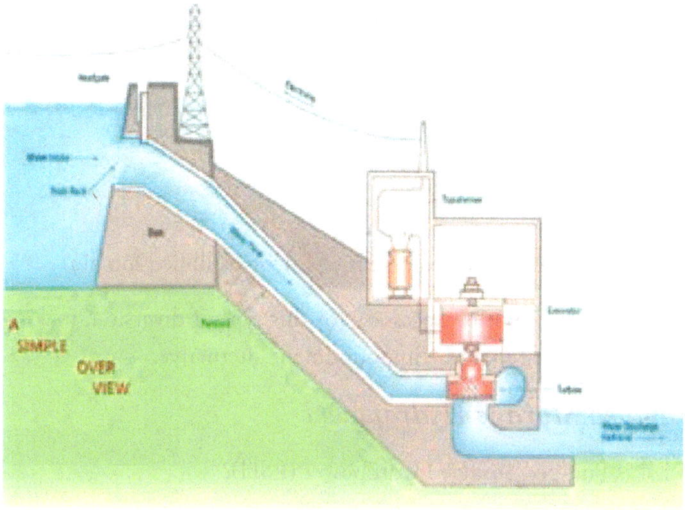

What can impoundment facilities do besides generating electricity?

It can provide water supply, control floods, and provide recreation in the form of boating, swimming, and fishing. The author, during his career as a water resources engineer, is familiar with these types of multi-purpose facilities.

What is a Run-of-River facility?

Run-of-river (aka Diversion) diverts a portion of a river flow using a diversion structure and a canal or channel to a forebay. Then, from the forebay, a penstock leads flow to a turbine with flows regulated by gates, valves, and turbines. The natural drop in elevation between the diversion structure and the tailrace where the diverted flow meets the river constitutes the net head that drives the turbines and generates the electricity. A typical schematic of the diversion is shown below:

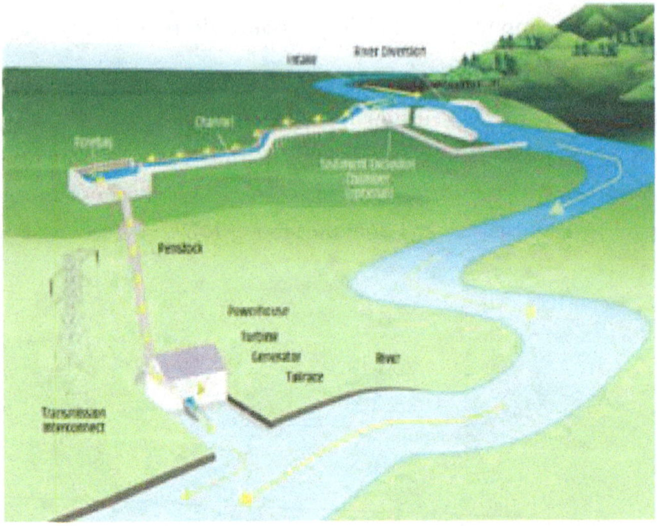

Source: U.S. Department of Energy (Public Domain)

A dam may not be necessary for this type of diversion, making it more economical than the impoundment alternative.

Pumped Storage Hydropower

What is Pumped Storage Hydropower (PSH)?

Briefly, it works like a battery – recharging it when the electricity demand is relatively low and discharging it when it is relatively high. So, it is an endless cycle of recharge – discharge. In practice, two water reservoirs are utilized in an electrical power grid system to make this work as follows:

1. Typically, there are two water reservoirs – one at a lower elevation (LEL) and one at a higher elevation (HEL).
2. When the electrical power grid system has a relatively low demand, water is pumped from the LEL reservoir to the HEL reservoir.
3. When the power grid system has a relatively high demand, water is released from the HEL reservoir to the LEL reservoir. The released

water flows by gravity and generates power by flowing through a turbine and generator, thus meeting the relatively high demand.

Following is a schematic of the workings of various components in a PSH (pumped storage) facility:

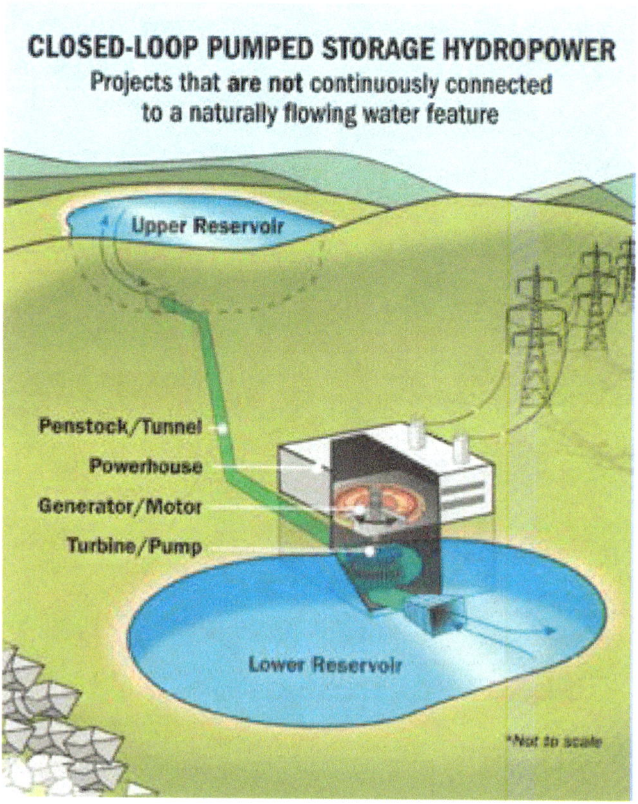

Source: U.S. Department of Energy (Public Domain).

An alternative to the above schematic is shown below:

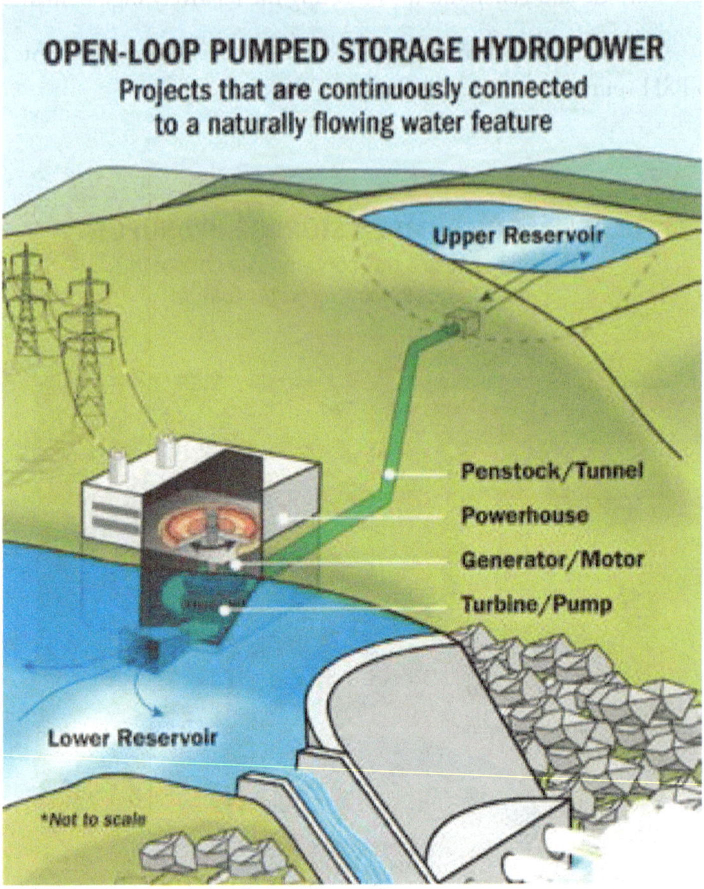

Source: U.S. Department of Energy (Public Domain)

While hydro depends on flowing water, another renewable resource, geothermal, taps into the Earth's inner heat.

Geothermal Energy

Why is geothermal energy considered to be renewable?

This is because the energy source is in the Earth's superheated core, which contains almost an inexhaustible supply of thermal energy. This thermal energy cools as it rises to Earth's crust, but it is available at relatively high temperatures.

What is the temperature gradient of thermal energy?

About 10,000 degrees Fahrenheit (just as hot as the surface of the Sun) at Earth's core is 4,000 to 3,220 miles below the Earth's surface.

About 900 degrees Fahrenheit, 10 miles below the Earth's surface.

About 50 - 60 degrees Fahrenheit 10 feet below the Earth's surface.

How does geothermal energy work?

1. When connected to a pipe, Hot water under pressure may reach the Earth's surface.
2. On reaching the surface, the hot water turns to steam due to ambient pressure.
3. Steam turns into a turbine connected to a generator and generates electricity.
4. (**Source:**USEPA.gov/public domain)
5. After generating electricity, the steam is cooled and condensed back into water, which is then returned underground to repeat the cycle.
6. The cooled water is pumped back into the Earth to begin the process again.

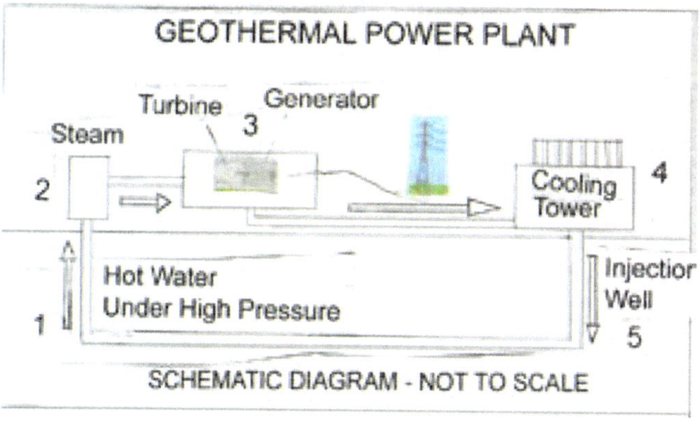

What are the advantages of geothermal energy?

As a source of renewable energy, its carbon dioxide emission is about one-sixth of the corresponding emission of a natural gas power plant.

It is about 80% cheaper than fossil fuels.

It is constantly available, unlike solar and wind, which depend on the sun shining and the wind blowing. (**Source**: TWI-global.com)

The largest drawback is that a geothermal plant needs to be built where geothermal energy is readily available, such as in Iceland. It may be more expensive for site investigation, leading to its sitting at other locations, such as in the western United States.

In which Earth regions are geothermal energy readily available?

Geothermal energies are readily available near major tectonic plate boundaries where most volcanoes are located, such as along the Ring of Fire which encircles the Pacific Ocean. The modern Theory of the motions of the Earth's outer layers is called plate tectonics.

It explains the existence of mountains, earthquakes, volcanoes, distribution of fossils, and ages of rocks. For example, when two tectonic plates collide at about 1.5 cm (0.6 inches) per year, they give rise to mountains, such as the Himalayas and the Alps. The Himalayas was formed about 50 million years ago, and the Alps was formed about 25 million years ago.

The Author could not verify these numbers independently and concludes that these are estimates. Mount Everest is the world's tallest mountain at 8,849 meters (or 29,032 ft) above sea level. Mount Everest began to form about 50 million years ago when the tectonic plate India rest on crashed into the Eurasian plate and then started to descend below it.

Picture credit: NASA.gov (Public Domain)

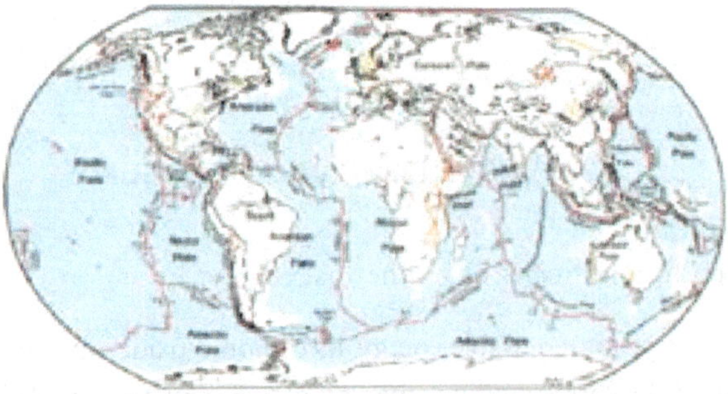

What are the Earth's tectonic plates?

Seven or eight major tectonic plates (aka lithosphere) and five minor tectonic plates (aka platelets) formed about 3 billion years ago. The schematic Global Tectonic Activity Map shows several faults, active spreading centers, and volcanoes. The geologists who prepared it achieved unprecedented completeness using historical data and satellite imagery.

In perspective, the Earth was formed 4.5 billion years ago when there were no plate tectonics, but there was only molten magma (very hot molten rocks). The Earth cooled down to form a crust.

Tidal Energy: Harnessing Tides To Generate Energy

Just as heat rises from Earth's core, the gravitational pull of the moon and sun stirs the seas, creating tides that can also be harnessed for power.

What creates tide in the first place?

The gravitational pull of the sun and the moon plus the rotation of the earth creates tides in coastal areas in a cyclical fashion: two rising tides (a.k.a. flood tides) and two falling tides (a.k.a. ebb tides) within about 24 hours 50 minutes. Flood tides occur every 12 hours and 25 minutes apart, and ebb tides occur every 12 and 25 minutes apart.

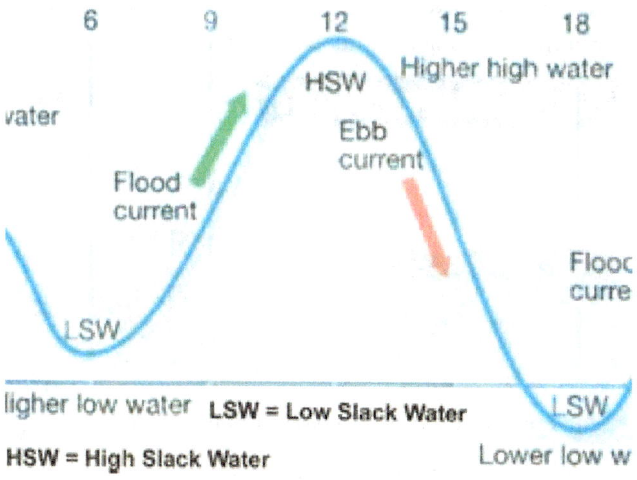

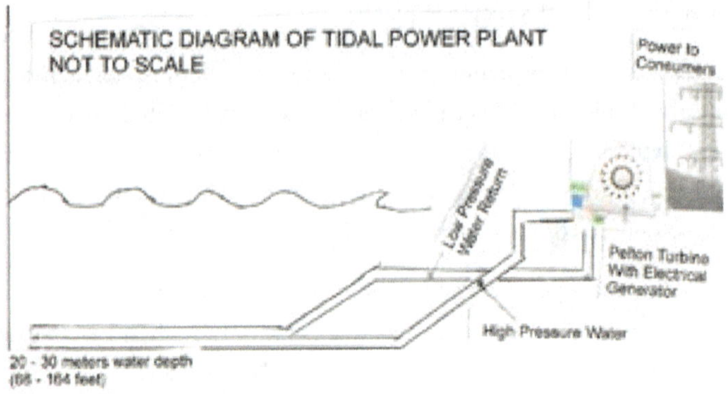

How is energy generated by the fluctuations of the flood current and the ebb current?

A Pelton turbine with an electrical generator is installed at the end of a high-pressure water line; the high tide causes the high pressure. The electricity generated is passed onto the consumers.

How is a man-made barrage system used to generate electricity?

A tidal barrage is similar to a dam installed across an inlet of an ocean that forms a tidal basin. Sluice gates on the barrage control water levels and flow rates to allow the tidal basin to fill on the incoming flow tide (aka high tide) and empty through a turbine that generates electricity on the ebb tide.

Alternatively, a two-way tidal power system generates electricity from both the incoming high tide and the outgoing ebb tide.

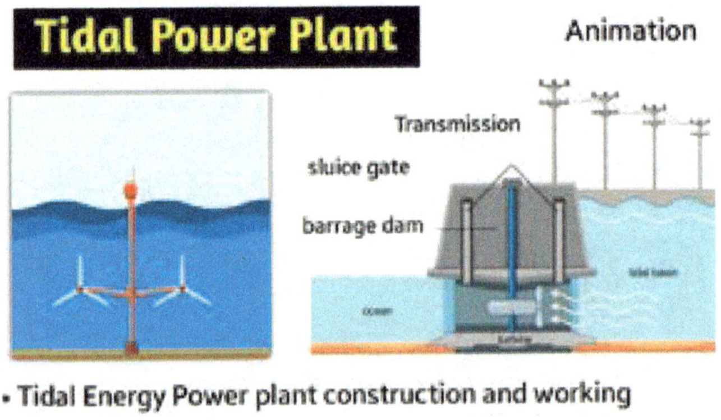

Source: U.S. Energy Information Administration (U.S. EIA)-Public Domain

What are the pros and cons of tidal energy?

Pros:

1. Tidal energy is a consistent and renewable source, unaffected by weather or time of day.
2. It has a high power output compared to other renewables.
3. It is predictable.
4. It has zero carbon emissions.

Cons:

1. Suitable tidal power sites are limited as these need to be located along a coastline and need to have a significant difference between high and low tides.
2. Expensive.
3. Power produced is usually out of sync with the peak power demand, requiring storage batteries.
4. Turbines can harm the surrounding ecosystem, including fish and other marine life.

Alongside tidal flows, the ocean's surface holds another powerful force: waves.

Wave Energy

What kind of energy-generating potential do waves have?

Ocean waves carry vast amounts of kinetic energy, making them a promising source of clean electricity. The theoretical annual energy potential of waves off the coasts of the United States is estimated to be 2.64 billion kilowatt hours, equal to about 63% of total U.S. utility-scale generation in 2023.

The west coasts of the United States, Europe, and the coasts of Japan and New Zealand have potential sites for harnessing wave energy. Many different methods and technologies for capturing and converting wave energy to electricity are under development. These methods include placing devices on or just below the water's surface and anchoring the devices to the ocean floor.

Source: U.S. Energy Information Administration (EEA)-Public Domain

What is wave energy?

The up and down (oscillatory) movement of waves creates energy which can be harnessed by wave energy converters (WECs) to produce electricity. Wave energy is renewable and non-polluting. A typical WEC may look as shown below:

What type of energy transfer takes place between the wind and the wave?

As the wind continues to flow across the surface of the body of water, its kinetic energy is transferred to the ocean's surface, making the wave amplitude bigger. These bigger waves are called gravity waves because their potential energy is due to the earth's gravitational force.

What are the pros and cons of wave energy?

Pros:

1. Zero emissions into the atmosphere.
2. Renewable form of energy.
3. Huge energy potential.

Cons:

1. Expensive.
2. The environment may be adversely affected.
3. Scalability issues: The project may not be readily expanded without additional expenses.

What are some forms of wave energy converters (WEC)?

1. These are wave profile devices shown below:

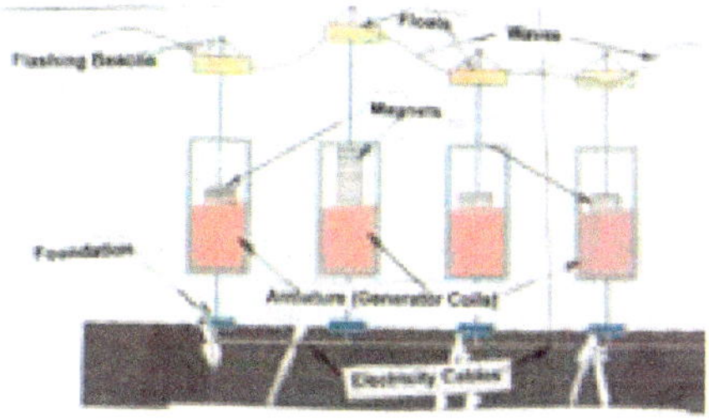

WAVE TO ELECTRICITY CONVERTER

……Source: U.S. Energy Information Administration (EIA)-Public Domain.

The up-and-down motion of the waves causes several floats (four shown above, in this instance) to move up and down. Each floater has magnets that move within an armature (aka generator coils). This movement induces an electrical current generated within the armature as predicted by Faraday's law. This electrical current is collected by cables and transferred to the consumers.

While wind, water, and sun are external forces, energy can also come from the organic materials we discard every day: biomass.

Biomass Energy

What is biomass energy?

Biomass energy is obtained by burning organic materials such as plant waste, animal waste, municipal solid waste (MSW), and forest waste, thereby producing heat and electricity and liquid fuel (ethanol and biodiesel) that can be used in automobiles, ships, and airplanes.

What are the pros of using biomass energy?

Biomass energy is a renewable source derived from organic materials such as plant, animal, and municipal solid waste. It is derived from recently living organic materials known as biomass, which can be used to produce transportation fuels, heat, electricity, and products such as plastics, lubricants, and industrial chemicals.

What are the cons of using biomass energy?

It emits greenhouse gases into the atmosphere. It requires lots of space. Compared to similar renewable sources of electricity, biomass is expensive to gather, transport, and store. It creates competition for arable land required for food and fiber production. The soil nutrients are depleted, and water quality is impaired.

What are the benefits of a robust bioenergy industry?

Abundant and renewable bioenergy can contribute to a more secure, sustainable, and economically sound future by:

1. Supplying domestic clean energy sources.
2. Reducing the U.S. dependence on foreign oil.
3. Generating U.S. jobs.
4. Revitalizing rural economies

The United States has the potential to produce 1 billion dry tons of non-food biomass resources annually by 2040 and still meet demands for food, feed, and fiber. One billion tons of biomass could:

1. Produce up to 50 billion gallons of biofuels.
2. Yield 50 billion pounds of bio-based chemicals and bioproducts.
3. Generate 85 billion kilowatt-hours of electricity to power 7 million households.
4. Contribute 11 million jobs to the U.S. economy.
5. Keep $200 billion in the United States.

Source: U.S. Department of Energy (DOE).

In addition to traditional renewables, emerging technologies like green hydrogen are opening new paths to a cleaner energy future.

Green Hydrogen Energy

What is green hydrogen energy?

Hydrogen is produced by electrolyzing water. However, the electricity must come from a renewable energy source to be considered green. Thus, the more renewable energy in the electricity-fuel mix, the "greener" the hydrogen produced.

Why is green hydrogen energy important?

Green hydrogen is a clean fuel. It emits only water vapor, and there is no carbon dioxide or other pollutants when used.

What are the advantages of green hydrogen?

Green hydrogen is an abundant, clean fuel that could power transportation and factories without emitting greenhouse gases.

What are the disadvantages of using hydrogen?

The process of producing hydrogen is very expensive. Moreover, if the energy that goes to produce hydrogen comes from burning coal, oil, or natural gas, it defeats the very purpose of "green" hydrogen.

Which country has the maximum number of green hydrogen plants?

As of 2024, Australia leads the world in the number of green hydrogen plants, reflecting its strong commitment to becoming a global exporter of clean energy. Backed by abundant solar and wind resources, Australia invests heavily in hydrogen to support domestic energy needs and international markets.

Several European countries are also at the forefront. Germany and Spain, each with 50 plants, use green hydrogen to decarbonize industry and transport, which aligns with EU climate targets. The Netherlands and the United Kingdom follow closely, integrating hydrogen into their broader energy transition plans.

In the United States, the federal government has increased support for hydrogen through infrastructure funding and tax incentives. With 37 plants, the U.S. expands hydrogen use in heavy transport and industrial sectors.

Other notable contributors include:

- Russia and Portugal, focusing on domestic energy applications.
- China, which, despite lower numbers, is scaling up rapidly.
- Chile, leveraging its renewable-rich landscape to enter the hydrogen mark

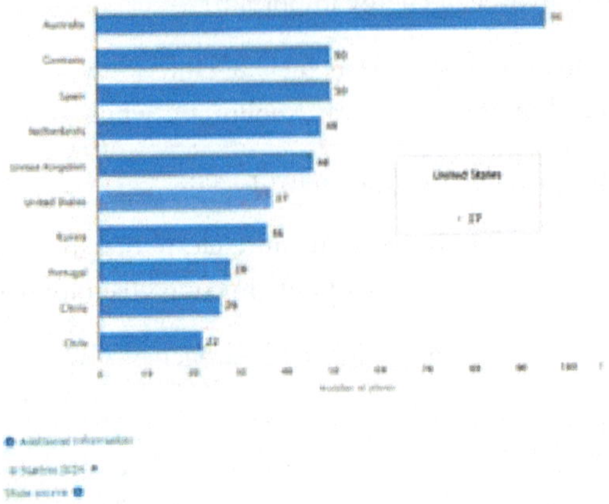

To sum up, while renewable energies offer hope for a sustainable future, reducing our carbon footprint also requires individual and collective responsibility.

Nuclear energy also offers a low-carbon path, though it raises its own set of questions. The next chapter explores its role in the fight against climate change.

CHAPTER 6
NUCLEAR ENERGY

Nuclear energy is a powerful source of electricity that produces almost no greenhouse gas emissions during operation.

While it is not considered renewable, it plays an important role in reducing carbon emissions worldwide. This chapter explores how nuclear energy works, its benefits, challenges, and past accidents that have shaped public perception.

Why is electrical power obtained from nuclear power plants considered clean power?

Nuclear power is considered clean because it produces electricity without releasing carbon dioxide or other greenhouse gases that are produced when burning fossil fuels.

Even though electricity obtained from nuclear power plants does not pollute the atmosphere, why it is not considered renewable energy?

Nuclear energy is not renewable because its main fuel, uranium, comes from limited underground deposits. These resources will eventually be exhausted.

How much uranium is available worldwide and how long will it last?

Globally, known uranium deposits total about 8.1 million tonnes. Another 9.3 million tonnes are considered "unconventional resources." Together, this supply could last around 230 years at current consumption rates.

Source: Nuclear Energy Agency (NEA), International Atomic Energy Agency (IAEA) – 2020 Report.

What are nuclear reactors?

A nuclear reactor is the core of a power plant. It controls nuclear chain reactions that release heat. This heat turns water into steam, which spins a turbine to generate electricity.

How do pressurized water reactors (PWRs) work?

More than 65% of the commercial reactors in the United States are pressurized water reactors (PWRs). These reactors pump water into the reactor core under high pressure to prevent water from boiling.

In a PWR, water flows through the reactor core under high pressure to absorb heat from nuclear fission. This hot water then enters a heat exchanger, where it heats a separate water supply to create steam. The steam powers a turbine to produce electricity. The original water is then cycled back into the reactor to be heated again.

PRESSURIZED WATER REACTOR (PWR)

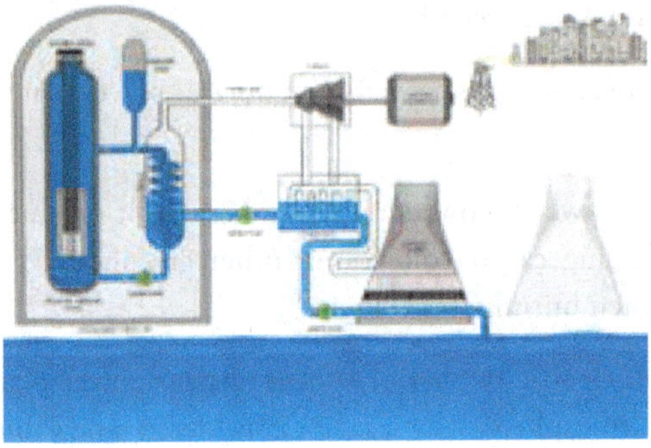

Image source: U.S. Department of Energy (Public domain).

What are the pros of nuclear energy?

1. Produces large amounts of carbon-free electricity.
2. Supports national power grids.
3. Has non-electric applications (e.g., medicine, space exploration).
4. Creates skilled jobs.
5. Promotes international cooperation on peaceful nuclear technology.

What are the cons of nuclear energy?

1. Public concern over safety and past accidents.
2. High costs for construction and maintenance.

3. Challenges in storing nuclear waste.

4. Misconceptions about links to nuclear weapons.

Which nuclear accidents caused global concern?

Fukushima, Japan nuclear accident of 2011. Fukushima is located in the Fukushima prefecture of Japan in northeastern part.

What happened at Fukushima power plant?

In 2011, a 9.0-magnitude earthquake triggered a 15-meter tsunami that disabled the cooling systems at three Fukushima Daiichi reactors in Japan. The loss of power caused a partial meltdown. One person later died from radiation-related illness, and over 2,200 deaths were linked to the evacuation process.

Where is Chernobyl located?

In 1986, Chernobyl was part of the Soviet Union. It is roughly 90 km (60 miles) north of Kyiv, Ukraine.

What happened at Chernobyl during April 25 – 26, 1986?

The 1986 Chernobyl disaster in present-day Ukraine was caused by human error during a poorly planned safety test. Operators disabled key safety systems, leading to an uncontrollable chain reaction. Explosions and a fire released large amounts of radiation into the atmosphere. Fatalities range from 2 to 50, with many more affected by radiation sickness.

What fatality counts resulted from the Chernobyl nuclear disaster?

Conflicting reports: as low as 2 fatalities to 50 fatalities. Dozens more contracted serious radiation sickness: some of them later died.

Where is Three Mile Island located?

Three Mile Island (TMI) is located near Harrisburg, Pennsylvania, U.S.A.

What happened at TMI in 1979?

In 1979, a mechanical failure at the Three Mile Island plant in Pennsylvania led to a partial meltdown. While no injuries or deaths occurred, the incident raised widespread concern and led to stricter safety regulations.

Were there any adverse effects due to this accident?

No injuries, deaths, or direct health effects occurred. No adverse effects occurred in the surrounding environment.

Source: U.S. DOE, NRC.

To sum up, nuclear energy remains a powerful tool in reducing carbon emissions, but it brings unique risks and challenges. Understanding its role in the broader energy mix helps us weigh its value alongside other clean energy options.

The next chapter turns to the question of carbon footprints and what individuals and nations can do to reduce them.

CHAPTER 7
ALL ABOUT CARBON

Carbon is a natural part of life on Earth, but human activities have dramatically increased the amount of carbon dioxide in the atmosphere.

According to the Columbia Climate School (New York, NY, U.S.), understanding the carbon cycle and how human activity alters it is key to addressing climate change.

This chapter explores how carbon is emitted, how we can reduce our individual and collective footprints, and what technologies are being developed to remove carbon from the environment.

What Are Carbon Emissions and Why Do They Matter?

Imagine Earth wearing a giant invisible blanket made of gases like carbon dioxide (CO_2) and methane (CH_4), which trap the sun's heat. Without it, the planet would be too cold to live on. But today, the blanket is getting too thick, and that's a problem.

Carbon emissions are what add layers to this blanket. They come mostly from burning coal, oil, and gas to power homes, cars, factories, and airplanes. Cutting down forests, raising livestock, and throwing waste into landfills also add carbon-based gases to the air.

The more carbon we release, the hotter the planet gets. To keep the Earth livable, we need to stop adding to the blanket and start finding ways to pull some of those layers off.

Fortunately, there's plenty we can do to reduce our carbon footprint.

What is a carbon footprint? It is a measure of the amount of carbon dioxide (CO_2) and other carbon compounds, such as methane (CH_4), emitted due to the consumption of fossil fuels by a person, community, country, etc.

How to Reduce Your Carbon Footprint

There are some easy ways to reduce your carbon footprint that do not require much effort or monetary investment.

Food

What you eat has a major impact on the environment. Animal products, especially beef and dairy, produce far more emissions than plant-based foods.

- **Eat less meat and dairy:** Skipping meat and dairy for one day can reduce your carbon footprint by about 8 pounds, nearly 2,920 pounds a year.
- **Buy local and organic**: Organic farming avoids synthetic fertilizers. Locally grown food reduces transport emissions.
- **Cut food waste:** Roughly one-third of all food produced globally is wasted. And wasted food means wasted water, energy, and land. (Source: United Nations Food and Agriculture Organization)

Clothes

The fashion industry accounts for 8–10% of global carbon emissions, more than all international flights and shipping combined. Most emissions come from production, transportation, and short product lifespans.

- **Avoid fast fashion:** Trend-based clothing is often low quality and quickly discarded.
- **Buy fewer, better items:** One high-quality shirt can replace several cheaper ones.
- **Wash less often and in cold water:** Heating water uses energy.
- **Air dry your clothes:** Dryers consume large amounts of electricity.
- **Donate, recycle, or repurpose:** Keep textiles out of landfills, where they emit methane.

Shopping

Everything we buy—electronics, plastic, packaging—comes with an environmental cost.

- Buy only what you need.
- Choose reusable products over single-use items.
- Recycle paper, plastic, metal, and glass.
- Repair rather than replace when possible.

Home

The way you use energy at home can either raise or reduce your footprint.

- Switch to LED bulbs and efficient appliances.
- Turn off lights and electronics when not in use.
- Unplug devices—they still use power when idle.
- Improve insulation to cut heating and cooling needs.

Transportation

How you get around plays a major role in your emissions.

- Walk, bike, or take public transport when possible.
- Carpool or combine errands to reduce trips.
- Maintain your vehicle—check tire pressure and drive smoothly.
- Choose electric or hybrid vehicles if available.

Air Travel

Airplanes release carbon at high altitudes, which amplifies their warming effect. Though only a small share of people fly regularly, aviation contributes about 2.5% of global CO_2 emissions, and that number is growing.

- **Fly less:** Avoid short-haul flights where trains or buses are available.
- **Choose nonstop flights:** Takeoffs and landings use the most fuel.
- **Fly economy:** Economy seats have a lower carbon footprint per passenger than business or first class.
- **Offset your flights:** Support carbon offset programs that invest in reforestation, clean cookstoves, or renewable energy. (Source: International Council on Clean Transportation; Our World in Data)

Carbon Offsets: (a.k.a. Carbon Credits)

A carbon offset is a payment that supports projects which reduce or remove carbon emissions, such as tree planting or renewable energy development.

If you offset one ton of carbon, the offset will help capture or destroy one ton of greenhouse gases that would have otherwise been released into the atmosphere. Offsets also promote sustainable development and increase the use of renewable energy.

Carbon Removal: Innovative Approaches

(Source: Investopedia, Patch Group)

What is carbon capture and storage (CCS)?

Carbon capture and storage is the process of separating carbon dioxide from emissions during industrial activities, such as power generation. The captured CO_2 is then stored deep underground, preventing it from entering the atmosphere and contributing to global warming.

What are the different technologies that are in use, or being developed for carbon capture?

These include the following:

1. Post-combustion carbon capture

This is the most widely used method today. It captures CO_2 from flue gases, emissions released after fuel is burned in industrial plants or power stations. In one technique called absorption, the gas is piped into a chamber called an absorber, where chemical solvents bind to the CO_2 and separate it from other gases. The CO_2 is then stripped from the solvent, which can be reused, and the CO_2 is compressed for transport and underground storage.

2. Pre-combustion capture

This method removes CO_2 from the fuel before it is burned. The fuel is partially oxidized, producing a mixture of hydrogen and carbon dioxide. The CO_2 is separated and captured, and the hydrogen can be used as a clean fuel.

3. Oxy-fuel combustion capture

Here, fuel is burned in nearly pure oxygen rather than air. This produces exhaust gas that is mostly CO_2 and water vapor. The water is condensed, and the remaining CO_2 is captured more easily than from regular flue gases.

4. Direct air capture (DAC)

Unlike the above methods, which work at emission sources, DAC pulls CO_2 directly from the air. Large fans draw air into a device called a collector, where CO_2 is extracted using chemical solutions or solid filters. The CO_2 is then stored or reused. As of 2024, DAC remains a developing technology with limited large-scale use.

5. Forestry

Trees naturally remove carbon through photosynthesis. Forestry-based solutions include reforestation, afforestation, and improved forest management (IFM) such as deferred harvesting. All are nature-based methods and rely on long-term carbon storage in trees and soil.

6. Ocean-related carbon removal

The ocean holds CO_2 in two main ways: as dissolved gas in water and as organic material like seaweed.

- **Seaweed sinking** involves growing kelp or algae and then sinking it more than 1,000 meters below the surface, where the carbon it contains stays trapped at the seafloor.
- **Ocean alkalinization** is an engineered method where substances like lime or olivine are added to seawater. These react with atmospheric CO_2, converting it into stable carbonate forms that stay locked in the ocean.

7. Mineralization

In mineralization, CO_2 reacts with natural or industrial minerals and forms solid carbonates.

- **Enhanced weathering** is one form, where reactive rocks are ground and spread across land or water.

- **Concrete injection is another:** captured CO_2 is added to wet concrete, where it binds with minerals, strengthening the concrete and storing the carbon. The concrete industry is responsible for about 8% of global CO_2 emissions, so this method holds strong potential for impact.

8. Biomass

Biomass refers to plant and animal material that contains carbon. While sometimes seen as emissions avoidance, some forms of biomass use result in long-term carbon storage. These include:

- **BECCS (Bioenergy with Carbon Capture and Storage):** Biomass is burned for energy, and the resulting CO_2 is captured and stored.

- **Bio-oil and biochar:** Biochar is a black, porous material similar to charcoal but produced at higher temperatures and from agricultural waste like corn husks. It stores stable carbon and is used to enrich soil. Though BECCS is engineered, biochar blends both natural and human processes. It also produces heat, which can support nearby farming or villages.

9. Soil

Soil is another natural carbon sink. Techniques such as regenerative agriculture, regenerative grazing, and soil amendments can enhance soil's ability to absorb and retain carbon.

- Cover cropping and reduced tillage help increase organic matter in soil.

- These practices can improve yields, reduce emissions from soil disturbance, and pull more carbon from the air into the ground.

Carbon Markets

What are carbon markets?

A carbon market is a special type of financial market where people buy and sell carbon credits in addition to stocks and bonds. Carbon credits are essentially permits that allow a purchaser to emit a certain amount of carbon dioxide or other greenhouse gases.

How much is one carbon credit?

One carbon credit allows the emission of one metric ton of carbon dioxide or the equivalent of other greenhouse gases. BTW, carbon credit is also known as carbon allowance.

What is the difference between carbon credits and carbon offsets?

Carbon credits generally represent a reduction in greenhouse gas emissions, whereas carbon offsets can represent removal of greenhouse gas.

What is the goal of the carbon credit system?

It is to reduce the emissions of greenhouse gases into the atmosphere.

Who are the key players in the carbon credit/ carbon offset market, and how do they operate?

Some national governments regulate and run the carbon market. These are:

1. European Union (EU) Emissions Trading System (ETS)
2. The California Global Warming Solutions Act (USA)
3. The China Certified Emissions Reduction (CCER) Scheme

Some industries are required to participate in carbon markets, while others may join voluntarily. Government agencies regulate emissions from sectors like refining, steel, transportation, power, and manufacturing.

Under carbon credit programs, companies in these regulated industries must reduce their greenhouse gas emissions, primarily carbon dioxide, or face penalties.

Are there some situations with industries whose participation is mandatory?

These industries can generate carbon credits, also known as offsets, by reducing or removing carbon dioxide. Reduction involves actions that lower emissions, such as installing solar panels or building wind farms.

Removal refers to projects that extract carbon dioxide from the atmosphere and store it, such as afforestation or carbon capture technologies. These credits can then be sold to other entities seeking to offset their emissions.

Reducing carbon emissions is essential to avoiding the most severe effects of climate change. Whether through daily choices, new technologies, or global policies, every action counts. In the next chapter, we turn to how world leaders come together to address climate change on a global scale.

CHAPTER 8

CARBON'S IMPACT ON EARTH'S ECOSYSTEMS

Carbon doesn't just stay in the air. Once released, it spreads across the planet: into oceans, forests, soil, and ice. These natural systems absorb, store, and sometimes release carbon, shaping the Earth's climate in the process.

But as emissions rise, the balance shifts. Oceans warm and turn acidic. Coral reefs bleach. Forests burn or are cut down. Ice melts faster than it forms. The very systems that once kept the planet stable now show signs of strain.

In this chapter, we explore what carbon does when it leaves the smokestack and enters the world, and how Earth's natural systems are reacting.

The Ozone Layer

The ocean is like a giant sponge and thermostat. It absorbs heat and carbon dioxide from the atmosphere. But as we add more carbon, the ocean gets warmer and more acidic, like a sponge reaching its limit and starting to overflow. This damages marine life, especially organisms like coral and shellfish.

Microscopic marine algae called phytoplankton, along with certain bacteria, are critical to the global carbon cycle. These tiny lifeforms absorb as much carbon as all the plants and trees on land combined, which makes the ocean one of Earth's largest carbon sinks.

Is ozone good or bad for global warming?

Think of the ozone layer as Earth's natural sunscreen. It blocks harmful ultraviolet (UV) rays from the sun. But when that sunscreen gets thin, more UV rays reach us, causing more skin cancer, eye problems, and ecosystem damage.

It is split into two zones:

- In the stratosphere, ozone is helpful. It protects us.
- In the troposphere, near the ground, ozone is a pollutant that harms the lungs and crops.

Let's explore these in detail.

What is stratospheric ozone?

Ozone is naturally produced in the stratosphere in a two-step process. In the first step, UV breaks apart an oxygen molecule (O2) to form two separate oxygen atoms (O + O). In the second step, each atom undergoes a binding with another oxygen molecule to form an ozone molecule (O+O2=O3).

Source: NOAA Chemical Sciences Laboratory (gov).

What is tropospheric ozone?

Near the earth's surface, the ozone layer is produced in chemical reactions involving naturally occurring gases and gases from pollution sources. Production reactions primarily involve hydrocarbon and nitrogen oxide gases and require sunlight. Fossil fuel combustion that produces greenhouse gases is a primary pollution source for tropospheric ozone production. The amount of surface ozone is relatively small compared to stratospheric ozone. As in the stratosphere, ozone in the troposphere is destroyed by naturally occurring chemical reactions and in reactions involving human-produced chemicals. Tropospheric ozone can also be destroyed when it reacts with a variety of surfaces, such as those of soils and plants.

At what elevations above the earth's surface stratospheric and tropospheric ozone can be found?

This is illustrated in the following image:

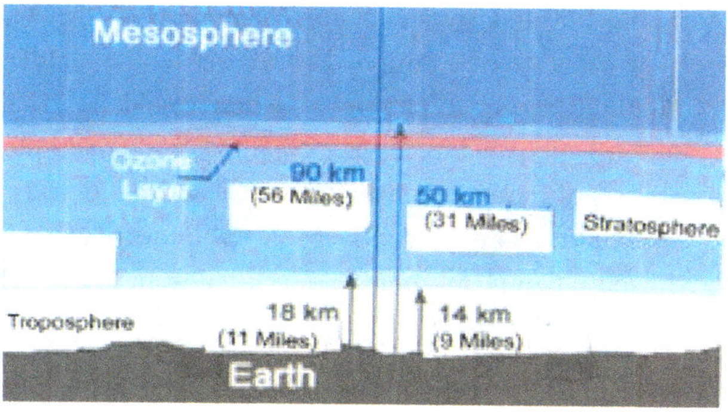

Image Source: National Science Foundation (Public Domain)

What is creating holes in the ozone layer in the stratosphere, particularly over Antarctica?

The ozone hole has developed because people have polluted the atmosphere with chemicals containing chlorine and bromine. The primary chemicals involved are chlorofluorocarbons (CFCs for short), halons, and carbon tetrachloride. These are known as ozone-depleting substances (ODS).

How did the Montreal Protocol address the ozone hole issue?

The Montreal Protocol, signed in 1987, is one of the most successful global environmental treaties. It phased out chemicals like CFCs that were damaging the ozone layer. As of 2024, the ozone hole is healing, and the protocol has prevented millions of cases of skin cancer. It's also helped slow global warming by cutting back on powerful greenhouse gases.

(Source: UN Environmental Programme or UNEP).

How successful has the Montreal Protocol been?

As of 2024, the parties to the protocol have phased out 98% of ODS globally compared to 1990 levels. Because most of these substances are potent greenhouse gases, the Montreal Protocol is also contributing significantly to the global climate system.

From 1990 to 2030, the treaty's control measures are estimated to have reduced greenhouse gas emissions by the equivalent of 135 gigatons of carbon dioxide, the equivalent of 11 gigatons a year.

What is expected of the Montreal Protocol over the long term?

With the full and sustained implementation of the Montreal Protocol, the ozone layer is projected to recover by the middle of the century. Without the treaty, ozone depletion would have increased tenfold by 2050 compared to current levels and resulted in millions of additional cases of melanoma, other cancers, and eye cataracts. It has been estimated, for example, that the Montreal Protocol will saving an estimated two million people each year by 2030 from skin cancer.

Under the Kigali Amendment, actions to limit the use of HFCs (hydrofluorocarbons, a greenhouse gas) are expected to prevent the emissions of up to 105 billion tons of carbon dioxide equivalent of greenhouse gases, helping to avoid up to 0.5 degrees Celsius of global temperature rise by 2300 – a truly unparalleled contribution to climate mitigation efforts, and the single largest contribution the world has made towards keeping the global temperature rise "well below" 2 degrees Celsius, a target agreed at the Paris Climate Conference of 2015.

(Source: UN Environmental Programme or UNEP).

Has the ozone hole in the stratosphere recovered as of 2024?

Projections from the UNEP suggest that the Antarctic ozone layer will recover to 1980 level by around 2066, with recovery in the rest of the world between 2040 and 2045.

Coral Reefs

Coral reefs are like underwater cities, bustling ecosystems that support 25% of marine life. They provide food, storm protection, and jobs for millions of people.

What does a typical coral reef look like?

The following are two typical pictures of a coral reef. They are often called the "rainforest of the sea."

Picture Credit: NASA (Public Domain)

Picture Credit.: NASA (Public Domain)

How does a coral reef come into being?

Reefs are formed of colonies of coral polyps held together by calcium carbonate.

What kind of animals live in coral reefs?

Coral reefs provide a habitat for a variety of marine life, including various sponges, oysters, clams, crabs, sea stars, sea urchins, and many species of fish.

What are some of the interesting facts about coral reefs?

1. Corals are animals.
2. Corals can be fluorescent.

3. Corals are plankton/small fish.

4. Corals can move.

5. Corals support 25 percent of ocean life.

What are the positive impacts of coral reefs on human life?

1. They act as a natural barrier for coastlines by dampening rough waves during storms.

2. They are the source of several cutting-edge medicines.

3. They generate billions of dollars for local economies through ecotourism and fisheries, while also serving as a critical food source for many island nations.

What is the impact of climate change on coral reefs?

Climate change leads to a warming of the ocean which causes thermal stress that contributes to coral bleaching and infectious disease. Sea level rise may lead to an increase in sedimentation for reefs located near land-based sources of sediment. Sedimentation runoff can lead to the smothering of coral.

What is coral bleaching?

When the water gets too hot, corals expel the algae that live in their tissues. It's like kicking out roommates who pay rent in food and color. Without the algae, the coral turns pale and weak, a process known as bleaching. If heat stress continues, the corals may die.

Bleaching occurs when coral polyps expel the zooxanthellae that live inside their tissue, causing the coral to turn white.

The zooxanthellae are photosynthetic, and as the water temperature rises, they begin to produce reactive oxygen species. This is toxic, so the coral expels the zooxanthellae.

Since the zooxanthellae produce the majority of coral coloration, the coral tissue becomes transparent, revealing the coral skeleton made of calcium carbonate.

Most bleached corals appear bright white, but some are blue, yellow, or pink due to the pigment-protein in the coral. (Source: Wikipedia)

Rainforests

Rainforests are Earth's green lungs. They absorb carbon dioxide and store it in roots, trunks, and leaves. Trees in the Amazon alone store about 48 billion tons of carbon.

Cutting them down is a double blow: releasing stored carbon and reducing the planet's ability to absorb new emissions.

What has been the effect of rainforests on the carbonization of the world's atmosphere?

The world's rainforests are a boon in slowing down the carbonization of the world's atmosphere. Trees in the rainforests not only absorb carbon dioxide from our air, but they also store carbon in their roots, leaves, and trunk.

Boreal Forests

Boreal forests stretch across the top of the world, like a green ring around the Arctic. These forests store enormous amounts of carbon in their cold, frozen soils.

They are dominated by species of spruce, fir, larch, pine, birch, and aspen. Precipitation is generally in the form of snow.

Which countries have predominantly large boreal forests?

Boreal forests cover large expanses of Canada, Alaska, Russia, and Scandinavia. Collectively, this is known as the circumboreal zone.

What are the effects of climate change on boreal forests?

Temperature increases will likely have an impact on carbon storage capacities of the circumboreal zone: as permafrost areas thaw and more areas of boreal forests burn, large amounts of greenhouse gases are likely to be released.

Can boreal forests provide ecosystem services?

Yes. Boreal forests can be a natural ally against rising emissions and slowing down climate change. Spanning 1.3 billion acres (2,031,250 square miles), the boreal forest is the earth's largest terrestrial carbon storehouse, storing 28 billion tons of carbon, or 11% of the world's total.

Apart from being a carbon sink, they provide the following services:

1. Water and soil processes.
2. Ecological processes that affect species range and biodiversity patterns.
3. Provides fishing, hunting, gathering, recreation, spiritual activities, medicines, and economic opportunities.
4. Hosts many species of flora and fauna and unique wildlife. More than 85 species of mammals such as wood bison, elk, moose, woodland caribou, grizzly, etc.

Polar Ice Caps

The Arctic and Antarctic are Earth's natural mirrors. They reflect sunlight and help keep the planet cool. But as temperatures rise, these ice sheets are melting faster.

What has been the effect of climate change on the polar ice caps?

Effect of climate changes on the polar ice caps in the Arctic, the Antarctic, and Greenland: A warmer climate affects ice sheets in several ways.

The most evident impact observed to date is a loss of ice due to warmer air and ocean waters. Warmer air causes the ice to melt more quickly and flow more rapidly to the sea, especially in the low-elevation regions near the edges of the ice sheets.

A distinction is made between Arctic and Antarctic ice caps. The Arctic is an ocean, covered by a thin layer of perennial sea ice and surrounded by land. ("Perennial refers to the oldest and thickest sea ice). Antarctica on the other hand, is a continent, covered by a very thick ice cap and surrounded by a rim of sea ice and the Southern Ocean.

Both the ice caps in the Arctic and the Antarctic are losing ice due to ongoing warming of the earth's surface and ocean. Meltwater coming from these ice sheets is responsible for about one-third of the global rise in sea level since 1993.

Since 1992, Greenland and Antarctica have both lost ice overall. Greenland has lost an average of about 175 billion metric tons of ice per year and Antarctica lost more than 90 billion metric tons per year.

The total amount of ice lost by Greenland and Antarctica from 1992 to 2020 was enough to raise sea levels worldwide by an average of three-quarters of an inch or more. By comparison, the global average sea level increased by about 3 inches overall during this period.

CHAPTER 9

MYTHS VERSUS FACTS ON CLIMATE CHANGE

The following discussions are courtesy of Dechen Tsering. She is the United Nations Environmental Programme (UNEP) Asia Pacific Director.

Myth #1: Climate change has always happened, so we should not worry about it.

The climate has been relatively stable since the last ice age, around 10,000 years ago, a stability that made human civilization possible.

Now, the Earth is warming faster than at any time in the past 2,000 years and is about 1.2 degrees Celsius hotter than it was in pre-industrial times. The last 10 years have been the warmest on record, with 2023 smashing global temperature records.

Other key climate-related indicators are also spiking. Ocean temperatures, sea levels and atmospheric concentrations of greenhouse gases are rising at record rates while sea ice and glaciers are retreating at alarming speeds.

Myth #2: Climate change is a natural process. It has nothing to do with people.

While climate change is a natural process, human activity is pushing it into overdrive. A landmark report from the Intergovernmental Panel on Climate Change (IPCC), which draws on the research of hundreds of leading climate scientists, found that humans have been responsible for almost all global warming over the past 200 years. Most of the warming has come from the burning of coal, oil, and gas. The combustion of these fossil fuels is flooding the atmosphere with greenhouse gases, which act like a blanket around the planet, trapping heat.

By measuring everything from ice cores to tree rings, scientists have been able to track concentrations of greenhouse gases. Carbon dioxide levels are at their highest in 2 million years, while two other greenhouse gases, methane and nitrous oxide, are at their highest in 800,000 years.

Myth #3: A couple of degrees of warming is not a big deal.

Small temperature rises can throw the world's delicate ecosystem into disarray, with dire implications for humans and other living things. The Paris Agreement on climate change aims to limit average global temperature rise to "well below" 2 degrees Celsius, and to preferably 1.5 degrees Celsius, since pre-industrial times.

Even that half-a-degree swing could make a massive difference. The IPCC found that at 2 deg. Celsius of warming, more than 2 billion people would regularly be exposed to extreme heat than they would at 1.5 deg. Celsius. The world would also lose twice as many plants and vertebrate species and three times as many insects. In some areas, crop yields would decrease by more than half, threatening food security.

"Every fraction of a degree of warming matters", says Tsering.

Warming of just half a degree could mean the difference between most coral reefs dying, and nearly all disappearing.

It would likely lead to the loss of other marine species, many of which are a critical source of protein for coastal communities.

Myth#4: An increase in cold snaps shows climate change is not real

This statement confuses weather and climate, which are two different things. Weather is the day-to-day atmospheric condition in a location and climate is the long-term weather conditions in a region. So, there could still be a cold snap while the general trend for the planet is warming.

Some experts also believe climate change could lead to longer and more intense cold in some places due to changes in wind patterns and atmospheric factors. One much-publicized paper found that rapid warming of the Arctic may have disrupted the swirling mass of cold air above the North Pole in 2021. This unleashed sub-zero temperatures as far south as Texas in the United States, causing billions of dollars in damages.

Myth #5: Scientists disagree on the cause of climate change

A 2021 study found that 99% of climate research links global warming to human activity, confirming earlier findings from 2013.

"The idea that there is no consensus is used by climate deniers to muddy the waters and sow the seeds of doubt", says Tsering. "But the scientific community agrees: the global warming we are facing is not natural. It is caused by humans."

Myth#6: It is too late to avert a catastrophe, so we might as well keep burning fossil fuels.

While the situation is dire, there is still a narrow window for humanity to avoid the worst of climate change.

UNEP's latest Emissions Gap Report found that by cutting greenhouse gas emissions by 42 percent by 2030, the world could limit global temperature rise to 1.5 deg. Celsius compared with pre-industrial levels.

A little math reveals that to reach that target, the world must reduce its annual emissions by 22 billion tons of carbon dioxide equivalent in less than seven years. That might seem like a lot. But by ramping up financing and focusing on low-carbon development in key transport, agriculture, and forestry, the world can get there.

"There is no question the task ahead of us is massive," Tsering says. "But we have the solutions we need to reduce emissions today and there is an opportunity to raise ambition in the new round of national climate action plans."

Myth#7: Climate models are unreliable

Climate models are often attacked by skeptics. But over decades, the IPCC has found them to consistently reflect reality. A 2020 University of California study looked at 17 past models and found 14 closely matched observed warming.

Myth#8: Humanity is inventive; we can just adapt to climate change.

Some countries and communities can adapt to rising temperatures, lower precipitations, and other impacts of climate change. But many cannot.

The world's developing countries collectively need between US $215 billion and US $387 billion per year to adapt to climate change, yet only have access to a fraction of that total, found UNEP's latest Adaptation Gap Report. Even wealthy nations will struggle to afford the cost of adaptation, which in some cases will require radical measures, such as displacing vulnerable communities, relocating vital infrastructure, or changing staple foods.

In many places, people are already facing hard limits on how much they can adapt. Small island developing states, for example, can only do so much to hold back the rising seas that threaten their existence.

Without significant action to lower greenhouse gas emissions, communities will reach these hard limits faster and begin to suffer irreparable damages from climate change, say experts.

CHAPTER 10

CLIMATE CHANGE: A RELIGIOUS POINT OF VIEW

Long before climate change entered public awareness, most religions taught that humans are stewards of the earth, responsible for protecting and preserving it.

Here is an overview of what the major world religions teach about the earth and humanity's role in preserving it:

1. Christianity

In Genesis 2:15, God instructs Adam to work and care for the Garden of Eden, establishing stewardship as a divine duty.

Here are some key passages from the Bible that reflect this understanding:

- Genesis 1:26-28: "Then God said, 'Let us make man in our image, according to our likeness; and let them rule over the fish of the sea and over the birds of the sky and over the cattle and over all the earth, and over every creeping thing that creeps on the earth.'"
- The earth is the Lord's and all its fullness – Psalm 24:1.

These verses teach that God has created everything and entrusted it to the care of humans. Other verses that reflect this understanding include Haggai 2:8, Psalm 89:11, Exodus 19, Psalm 8:6-8, and Psalm 115:16.

2. Judaism

Judaism also affirms the command to care for creation, as seen in Genesis 1:26. This stewardship extends to all species. The Torah also specifies that humans take care of other species.

Most, if not all, Jews accept the science of climate change and global warming. In practical terms, Israel has reduced carbon dioxide (CO_2) emissions by operating clean-energy municipal buses and limiting new car emissions.

In fact, a ban on gasoline-powered cars will be imposed by 2030.

3. Buddhism

Buddhists advocate living in love and harmony with one another, which extends to all of nature. Buddhism also teaches compassion, simplicity, and mindful living. Practicing mindfulness can reduce consumer habits and one's carbon footprint.

4. Hinduism

Moving on, many Hindus strive to reduce and recycle as an expression of Dharma, right living that includes care for the Earth.

5. Islam

Islam also teaches that God (Allah) has made people stewards of the Earth with the responsibility to protect and preserve the environment.

6. Daoism (aka Taoism)

The Daoists believe that "Heaven and Earth and I live together." Thus, they try not to interfere with natural laws, which is instrumental in curbing the effects of climate change.

7. Shintoism

Shintoism teaches that the natural world is part of *Kami,* the Shinto God, which is an integral part of natural elements such as rivers, mountains, and trees. This connection between the divine and the natural world encourages people to be stewards of the natural world.

8. Confucianism

Confucianists subscribe to the "Golden Rule:" Do not do to others what you don't wish done to yourself. This rule then extends to animals, nature, and the Earth's ecosystem.

9. Sikhism

Sikhism views nature as a sacred gift from the Creator. The Guru Granth Sahib, the central scripture of Sikhism, describes the Earth as a "Dharamsaal," a place of spiritual practice.

Sikhs are also taught to live honestly, share with others, and care for creation. Wastefulness and greed are discouraged, while humility and sustainability are seen as signs of spiritual awareness.

10. Jainism

Finally, Jainism teaches *Ahimsa* (non-violence) as its highest principle. This includes not just people and animals but also plants, air, water, and soil.

Jains believe that every part of nature has a soul and must be treated with care and respect. Many Jains follow strict vegetarian or vegan diets and avoid activities that harm ecosystems.

This was a brief overview of how different religions approach the topic of climate change.

Hence, it is not just a scientific issue, but also a moral and spiritual one. What is encouraging is that religious teachings across traditions share a call to care for creation.

CHAPTER 11

CONFERENCE OF PARTIES (COP)

Since 1995, world leaders have gathered annually at the Conference of the Parties (COP) to discuss, negotiate, and shape global climate policy.

These meetings are held under the United Nations Framework Convention on Climate Change (UNFCCC) and serve as the main platform for tracking progress, setting emission targets, and funding climate solutions.

This chapter presents key developments from each COP, showing how global climate dialogue has shifted over time, arranged thematically.

SUMMARY OF COP MEETINGS AS OF DECEMBER 2023

	VENUE	DATES	
29	Baku, Azerbaijan	Nov. 11 - 22, 2024	
28	Dubai	12 Dec., 2023	
27	Sharm El Sheikh, Egypt	Nov. 6-12, 2022	
26	Glasgow, Scotland	31 Oct. - 12 Nov., 2021	SEE NOTE 1
25	Chile and Spain	25 Nov. - 1 Dec., 2019	
24	Katowice, Poland	2 Dec. - 14 Dec., 2018	
23	Bonn, Germany	6 Nov. - 17 Nov., 2017	
22	Marrakesh, Morocco	7 Nov. - 18 Nov., 2016	
21	Paris, France	30 Nov. - 12 Dec., 2015	a.k.a "The Paris Agreement".
20	Lima, Peru	1 Dec. - 12 Dec., 2014	

19	Warsaw, Poland	11 Nov. - 22 Nov., 2013	
18	Doha, Qatar	26 Nov. - 8 Dec., 2012	
17	Durban, South Africa	28 Nov. - 9 Dec., 2011	
16	Cancun, Mexico	29 Nov. - 10 Dec., 2010	
15	Copenhagen, Denmark	Dec. 7 - 18, 2009	
14	Poznan, Poland	Dec. 1 - 12, 2008	
13	Bali, Indonesia	Dec. 3 - 14, 2007	
12	Nairobi, Kenya	Nov. 6 - 17, 2006	
11	Montreal, Canada	28 Nov. - 9 Dec., 2005	
10	Buenos Aires, Argentina	Dec. 6 - 17, 2004	
9	Milan, Italy	Dec. 1 - 12, 2003	
8	New Delhi, India	Oct. 23 - Nov. 1, 2002	
7	Marrakesh, Morocco	Oct. 29 - Nov. 10, 2001	
6	The Hague, the Netherlands	Nov. 13 - 25, 2000	
5	Bonn, Germany	Oct. 25 - Nov. 5, 1999	
4	Buenos Aires, Argentina	Nov. 2 - 13, 1998	
3	Kyoto, Japan	Dec. 1 - 10, 1997	a.k.a "The Kyoto Protocol"
2	Geneva, Switzerland	Jul. 8 - 19, 1996	
1	Berlin, Germany	Mar. 28 - Apr. 7, 1995	

NOTE 1: COP26 was originally scheduled to take place
from 9 to 19 November 2020 in Glasgow, Scotland, but was
postponed from 31 October to 12 November 2021 due
to the COVID-19 pandemic.

Laying the Groundwork (COP 1–15)

This period focused on defining climate goals, legal frameworks, and early implementation strategies. It laid the foundations for international cooperation, including the Kyoto Protocol.

COP 1 – Berlin, Germany (1995)

Delegates agreed that the existing commitments under the UNFCCC were inadequate and needed strengthening. This led to the "Berlin Mandate," setting the stage for legally binding targets.

COP 2 – Geneva, Switzerland (1996)

The Geneva Ministerial Declaration endorsed the findings of the IPCC's Second Assessment Report and affirmed the scientific basis for stronger action on climate change.

COP 3 – Kyoto, Japan (1997)

The Kyoto Protocol was adopted. It set legally binding emission reduction targets for developed countries and introduced market-based mechanisms like emissions trading and the Clean Development Mechanism (CDM).

COP 4 – Buenos Aires, Argentina (1998)

The "Buenos Aires Plan of Action" was agreed upon, setting a timeline to finalize the operational details of the Kyoto Protocol.

COP 5 – Bonn, Germany (1999)

This was a technical session focusing on strengthening guidelines and reporting systems under the Kyoto framework. No major political decisions were made.

COP 6 – The Hague, Netherlands (2000)

Talks collapsed over disagreements on carbon sinks and compliance. A follow-up meeting was held in 2001.

COP 6 (Part II) – Bonn, Germany (2001)

Delegates salvaged the Kyoto Protocol by agreeing on the Bonn Agreements, which outlined rules for emissions trading, compliance, and financial mechanisms.

COP 7 – Marrakech, Morocco (2001)

The Marrakech Accords were adopted, finalizing operational rules for the Kyoto Protocol, including accounting procedures and reporting requirements.

COP 8 – New Delhi, India (2002)

Focused on adaptation and sustainable development, particularly for developing countries. Discussions highlighted equity and common but differentiated responsibilities.

COP 9 – Milan, Italy (2003)

Delegates clarified rules for the Clean Development Mechanism (CDM) and finalized guidelines for land use, land-use change, and forestry (LULUCF) projects.

COP 10 – Buenos Aires, Argentina (2004)

The conference reviewed progress under the UNFCCC and began informal talks on future commitments beyond 2012, the end of Kyoto's first phase.

COP 11 – Montreal, Canada (2005)

The Kyoto Protocol entered into force earlier that year. This COP launched formal discussions on a post-2012 framework, initiating the "Montreal Action Plan."

COP 12 – Nairobi, Kenya (2006)

Discussions focused on adaptation funding and support for African countries. Progress remained slow on long-term emissions targets.

COP 13 – Bali, Indonesia (2007)

The "Bali Road Map" was adopted, launching a two-year negotiation process toward a future climate agreement involving all countries, not just developed ones.

COP 14 – Poznań, Poland (2008)

Technical groundwork was laid for the Copenhagen conference. Discussions continued on adaptation funds and deforestation reduction.

COP 15 – Copenhagen, Denmark (2009)

The much-anticipated Copenhagen Summit ended without a legally binding agreement. However, the "Copenhagen Accord" recognized the goal of limiting global temperature rise to below 2°C.

The Paris Pivot (COP 16–22)

After Copenhagen's disappointment, countries shifted from top-down mandates to bottom-up commitments. This period culminated in the Paris Agreement, a global turning point in climate diplomacy.

COP 16 – Cancún, Mexico (2010)

The Cancún Agreements formalized climate financing pledges and set up the Green Climate Fund (GCF). They also established guidelines for transparency and created frameworks for adaptation and REDD+ (Reducing Emissions from Deforestation and Forest Degradation).

COP 17 – Durban, South Africa (2011)

Delegates launched the Durban Platform for Enhanced Action, which called for a new legal agreement to involve both developed and developing countries. This set the groundwork for what would become the Paris Agreement.

COP 18 – Doha, Qatar (2012)

The Kyoto Protocol was extended through 2020 with a second commitment period. However, several major emitters—like Canada, Russia, and Japan—did not join the extension.

COP 19 – Warsaw, Poland (2013)

Countries agreed on the "Warsaw International Mechanism" to address loss and damage from climate change impacts. The meeting also laid the groundwork for nations to prepare their own climate plans (INDCS).

COP 20 – Lima, Peru (2014)

Delegates agreed on a draft text for the Paris Agreement. Nations submitted their Intended Nationally Determined Contributions (INDCs), setting the stage for 2015.

COP 21 – Paris, France (30 November – 12 December 2015)

The landmark Paris Agreement was adopted. It aimed to limit global temperature rise to "well below 2°C," with efforts to stay under 1.5°C. All countries were required to submit updated and more ambitious national plans every five years.

COP 22 – Marrakech, Morocco (2016)

Known as the "Action COP," this meeting focused on turning the Paris Agreement into a working plan. Countries began developing implementation rules and continued negotiating finance and transparency mechanisms.

From Talk to Tension (COP 23–28)

After Paris, expectations rose. But so did frustration over finance, fossil fuel phaseouts, and implementation gaps. This period saw bold pledges, climate justice debates, and mounting pressure for accountability.

COP 23 – Bonn, Germany (2017)

Hosted by Fiji and held in Bonn, this COP focused on loss and damage, climate resilience, and transparency. Countries began fleshing out the "Paris Rulebook" to monitor and report progress.

COP 24 – Katowice, Poland (2018)

The "Katowice Rulebook" was adopted, establishing guidelines for how countries should track and report their emissions and climate actions. However, debates over equity and finance continued.

COP 25 – Madrid, Spain (2019)

Originally planned for Chile, this COP struggled to finalize rules for carbon markets under Article 6 of the Paris Agreement. The meeting ended with little progress, despite increasing public pressure.

COP 26 – Glasgow, UK (2021)

After a COVID-19 delay, Glasgow brought new energy. Countries agreed to "phase down" unabated coal and phase out fossil fuel subsidies. Over 140 nations submitted updated NDCs, and rules for carbon markets were finalized. Still, funding promises remained unmet.

COP 27 – Sharm el-Sheikh, Egypt (2022)

A major breakthrough came with the agreement to establish a Loss and Damage Fund to support vulnerable countries. However, the conference drew criticism for failing to advance fossil fuel phaseout language.

COP 28 – Dubai, UAE (2023)

For the first time, the final agreement explicitly called for a transition away from fossil fuels, marking a historic shift in COP language. Delegates also adopted the first "Global Stocktake," assessing collective progress toward the Paris goals. Still, some nations opposed stronger targets, and disagreements over climate finance persisted.

As climate agreements expanded, so did the need for funding, especially to support low-income countries facing the greatest climate risks. One major channel for this is the Global Environment Facility (GEF), established in 1991 to help nations meet environmental goals.

What is GEF (Global Environment Facility)?

It is a fund to basically recoup losses due to biodiversity loss, climate change, pollution, supporting land and ocean health. This financing helps developing countries to address their issues.

The partnership is all inclusive involving 186 governments, civil society, indigenous people, women and youth. GEF has provided $25 billion for financing and mobilized another $138 billion for country-driven priority projects.

The family of funds include the following: Global Environment Facility Trust Fund, Global Biodiversity Framework Fund (GBFF), Least Developed Countries Fund (LDCF), Special Climate Change Fund (SCCF), Nagoya Protocol Implementation Fund (NPIF), and Capacity Building Initiative for Transparence Trust Fund (CBIT).

To sum up, across nearly three decades, the COP process has brought together nearly every nation on Earth to confront a common threat.

These meetings have led to key achievements, from recognizing climate science to forming legally binding agreements and launching climate finance facilities.

Still, the gap between promises and implementation remains wide. The challenge now is not only to agree, but to act, and to act fast enough to avoid the worst outcomes of climate change.

CHAPTER 12
CLIMATE ACTIVISTS

Climate change has sparked a global movement unlike any before it. From school strikes to sit-ins, from legal challenges to shareholder revolts, individuals and groups across every continent are calling for urgent action.

Many activists are scientists, but countless other students, politicians, or ordinary citizens refuse to remain silent. While their tactics differ, all activists share a common belief: the climate crisis demands an immediate and bold response.

This chapter introduces key voices in the climate movement, the actions they have taken, and the challenges they continue to face.

Who Are Climate Activists?

Climate activists come from all walks of life. What unites them is their commitment to protecting the planet for future generations.

Here are a few of the most influential voices:

Greta Thunberg

This Swedish teenager has sparked a global movement by staging school strikes for the climate. Her blunt, science-driven speeches challenged world leaders and helped galvanize a new generation of activists.

Jane Fonda

Legendary Hollywood actress and long-time activist who launched "Fire Drill Fridays" to raise awareness about climate inaction. She also founded the Climate PAC (Political Action Committee) to support political candidates prioritizing the environment.

Sir David Attenborough

A British naturalist and broadcaster whose documentaries brought the wonders—and fragility—of nature into millions of homes. Over the years, he became a strong advocate for climate action, warning of irreversible damage.

Vanessa Nakate

A Ugandan climate activist who campaigns for both environmental action and social justice. She emphasizes that climate change hits poorer nations hardest, even though they contribute least to the problem.

Xiye Bastida

A Mexican-Indigenous youth activist and organizer with "Fridays for Future" and the "People's Climate Movement." She brings attention to how climate change affects Indigenous communities and marginalized groups.

Bill McKibben

An American environmentalist and founder of 350.org, one of the first large-scale grassroots movements to tackle climate change. He has been a major voice for fossil fuel divestment.

Among these, Greta Thunberg is arguably the most popular climate change activist today. Her speeches and interviews have captured global attention because of their clarity, urgency, and refusal to soften uncomfortable truths. Here are some of her most powerful statements, and why they matter:

- "I want you to act as if the house is on fire. Because it is."

Greta's signature phrase calls for immediate and sweeping action, not cautious long-term planning.

- "Our house is falling apart, and our leaders need to start acting accordingly."

She reminds politicians that the climate crisis is not a future problem; it is unfolding now.

- "You have stolen my dreams and my childhood with your empty words."

In her UN address, Greta directly confronted world leaders for making promises without meaningful action, speaking for millions of young people who fear for their future.

- "Hope is not passive. Hope is taking action."

Greta redefines hope, not as waiting for change, but as building it through persistence and courage.

- "The science is clear. We need drastic, immediate action."

Throughout her activism, Greta has grounded her message in scientific consensus, urging leaders to listen to experts rather than lobbyists.

- "We cannot solve a crisis without treating it like a crisis."

She criticizes half-measures and weak agreements, insisting that governments must respond to climate change with the urgency of a true emergency.

That said, climate change activism is more than just speeches and quotes.

Actions Taken by Climate Activists

As stated above, climate activists use a range of tactics to push for change. Some actions target governments, while others focus on corporations or the public. Their goal is to raise awareness, pressure decision-makers, and shift society toward a more sustainable future.

Here are some of the major types of actions:

- **School Strikes**

Students around the world, inspired by Thunberg, skipped classes on Fridays to demand stronger climate policies.

- **Public Protests and Marches**

Millions have marched in cities across the globe, making climate change visible to media, politicians, and voters.

- **Direct Action and Civil Disobedience**

Groups like Extinction Rebellion stage sit-ins, roadblocks, and nonviolent protests to disrupt business as usual and demand urgent action.

- **Court Cases and Legal Challenges**

Activists have taken governments and corporations to court, arguing that failure to act on climate change violates human rights and environmental laws.

- **Shareholder Activism**

Investors and activists pressure major corporations to disclose their carbon emissions, commit to net-zero targets, and divest from fossil fuels.

- **Online Campaigns and Awareness Drives**

Social media platforms amplify climate messages, allowing young activists, Indigenous leaders, and scientists to reach global audiences directly.

Here are some concrete examples of activism in action.

Recent Movements and Protests

Around the world, activists are organizing direct actions, political campaigns, and community initiatives to drive change from the ground up.

1. Dutch Extinction Rebellion Protesters

In 2023, thousands of climate activists in the Netherlands blocked highways and occupied public spaces, demanding an end to fossil fuel subsidies. Despite police crackdowns and mass arrests, the protests drew widespread public support and forced the Dutch government to debate subsidy reforms.

2. Ugandan Climate Strikers

Youth activists in Uganda, led by figures like Vanessa Nakate, organized school strikes and public rallies to highlight the disproportionate impact of climate change on Africa. Their message: countries least responsible for emissions are suffering the most from their effects.

3. Jane Fonda's Climate PAC

In the United States, actress and activist Jane Fonda launched the aforementioned Climate PAC to support political candidates who prioritize strong climate policies. Through Fire Drill Fridays and other campaigns, she brought climate activism into mainstream media attention and linked it to broader social justice movements.

Each of these movements shows that climate activism is no longer confined to scientific conferences or UN negotiations. It is visible, urgent, and deeply rooted in communities demanding real change. But it is not that straightforward.

Frustrations and Obstacles

While climate activism has made the climate crisis impossible to ignore, real progress often runs into powerful obstacles.

Government Inaction

Many leaders make bold promises at international summits but fail to deliver at home. Political cycles favor short-term goals over long-term climate strategies, leaving activists frustrated by broken pledges and watered-down legislation.

Corporate Lobbying

Fossil fuel companies and other major industries spend billions lobbying against climate regulations. In some countries, corporate influence shapes public policy more than scientific warnings do, slowing the transition to clean energy.

Misinformation and Denial

Despite overwhelming scientific evidence, climate misinformation continues to spread online and through some media outlets. This confusion undermines public support for necessary reforms.

Political Pushback

Some political leaders openly dismiss climate concerns. In the United States, former President Donald Trump rolled back major environmental protections and withdrew the country from the Paris Agreement during his term, signaling to others that climate action was optional.

Economic Inequality

Poorer nations and marginalized communities suffer the most from climate change but often have the least political power to drive solutions. Activists must constantly fight to keep these voices centered in climate conversations.

Despite these barriers, the climate movement persists. Activists continue to demand accountability, push for systemic changes, and remind leaders that science, not politics, should guide climate action.

Progress is often slow, and setbacks are common. Yet persistent activism pressurizes leaders to act, keeps public attention focused, and inspires new generations to join the cause. You are invited to do the same.

CHAPTER 13

THE REPORT CARD

After years of negotiations, promises, and plans, the real question remains: Are we making progress?

This chapter takes a hard look at where the world stands today on global temperatures, sea level rise, and the drive toward net-zero emissions.

Using the latest available data from trusted scientific agencies, we measure how far we have come and how far we still need to go.

What is the latest (2025) report on global warming?

According to the World Meteorological Organization (WMO) and the Copernicus Climate Change Service, the Earth's average surface temperature has increased by about 1.45 degrees Celsius compared to pre-industrial levels (1850–1900). This is dangerously close to the 1.5°C limit set in the Paris Agreement. Some years, like 2023, have seen temporary spikes reaching or slightly exceeding 1.5°C, although long-term averages remain just below that threshold.

Has global warming been verified by other independent agencies?

Indeed, it has. For instance, Copernicus Climate Change Services found in 2024 that global surface temperature has exceeded 1.5^{0}C, compared to the pre-industrial levels of 1850-1900. Here's the graph of the annual average global temperature increase between 1967 and 2024:

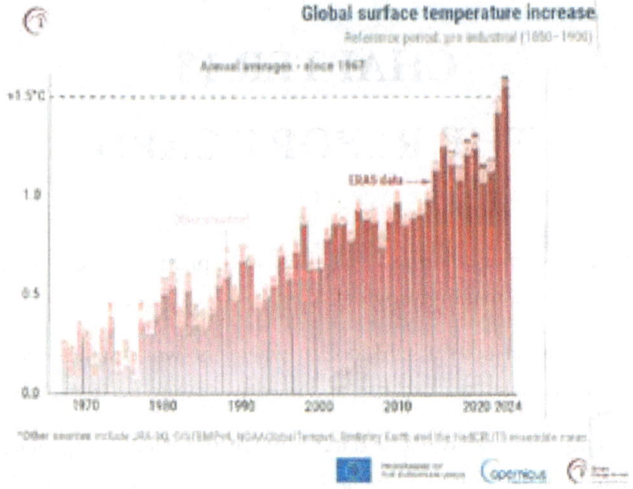

2024 is seen to have exceeded 1.50°C, which violates the Paris Agreement of 2015 (see COP 21). 2024 also happens to be the warmest on record.

What is the 2024 report on sea level rise?

Sea level rise is the increase in the global average level of the world's oceans, mainly caused by the melting of glaciers and polar ice sheets, and the expansion of seawater as it warms.

According to NASA, the global mean sea level has risen by about 8–9 inches (20–23 centimeters) since 1880. The rate of rise has accelerated sharply in recent decades, now averaging about 3.3 millimeters per year, based on satellite measurements.

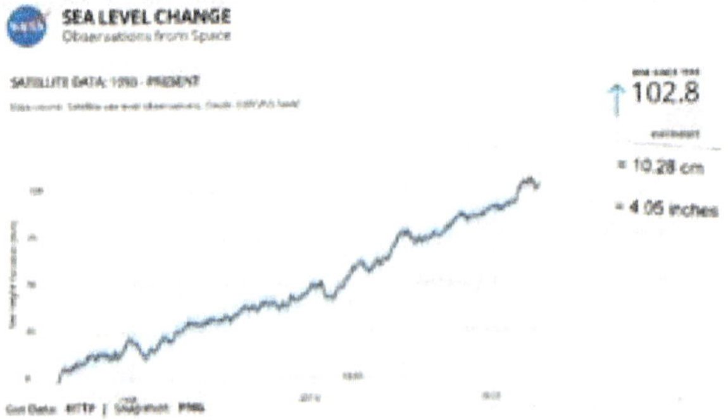

By 2100, scientists project sea levels could rise between 0.3 meters (about 1 foot) and 1.1 meters (about 3.6 feet), depending on global emissions. A rise of around 0.5 meters would already threaten millions living in coastal regions, while higher rises would make some islands and cities uninhabitable without major defenses.

What is Net Zero?

Net zero means balancing the amount of greenhouse gases released into the atmosphere with the amount removed. To reach net zero, emissions must be reduced as much as possible, and any remaining emissions must be offset through methods like carbon capture, tree planting, or other carbon removal technologies. Achieving global net zero by mid-century is seen as essential to limiting warming to 1.5°C.

Which countries have committed to Net Zero goals?

Many countries have made net-zero pledges, although timelines and legal backing vary. Some key examples include:

- **European Union**: Net zero by 2050 (European Green Deal, launched in 2019)
- **United States**: Net zero by 2050 (Executive orders and federal plans)
- **United Kingdom**: Net zero by 2050 (legally binding, passed in 2019)
- **China**: Net zero by 2060 (announced in 2020)
- **India**: Net zero by 2070 (announced at COP 26 in 2021)

However, many of these pledges are still in early stages of implementation, and interim targets (e.g., 2030 emission cuts) are often insufficient to stay on track for 1.5°C.

COUNTRY	2024 EMISSION STATUS	IN LAW	PROPOSED LEGISLATION	IN POLICY DOCUMENT	TARGET UNDER DISCUSSION	YEAR OF TARGETED NET ZERO EMISSIONS
BHUTAN						Achieved
SURINAME						Achieved 2014
NORWAY		X				2030
SWEDEN		X				2045
UNITED KINGDOM			Passed net zero emissions law in 2019			2050
FRANCE			Passed net zero emissions law in 2019			2050
SPAIN			Policy Position (2020)			2050
NEW ZEALAND			Climate Change Response (Zero Carbon) Amendment Act (2019)			2050
URUGUAY			Submission to UN (2020)			2030
FINLAND			Coalition Agreement (2019)			2035
IRELAND			Coalition Agreement (2020)			2050
DENMARK			Climate Act (2020)			2050
CHILE		Policy Position (2019) and Submission to UN (2020)				2050
PORTUGAL			Policy Position (2018)			2050
COSTA RICA		Policy Position (2019) and Submission to UN (2020)				2050
FIJI			Submission to UN (2020)			2050
MARSHALL ISLANDS			Submission to UN (2020)			2050
EUROPEAN UNION			Green Deal (2010)			2050
GERMANY			Climate Action Law (2020)			2050
THE NETHERLANDS			Climate Act (2019)		49%by 2030,95%by2050.100% clean electricity by 2050	
CHINA	Announcement to UN (2020) (2019)					2060
BRAZIL	Submission to UN (2020)					2060

The climate report card shows both progress and failure. The next few years will determine whether today's promises become real change, or whether the gap between words and action grows even wider.

CONCLUSION

Throughout this book, we have examined the causes and impacts of climate change, the role of key agreements, the activism it has inspired, and the measurable progress—or lack of it—that has followed.

The facts show that while many nations have made pledges toward reducing emissions and achieving net-zero goals, the current trajectory is still insufficient to meet the critical targets needed to limit global warming to 1.5°C.

The challenge ahead is steep but not insurmountable. Reducing carbon emissions, protecting natural systems, developing cleaner technologies, and fostering international cooperation remain essential. Climate change is a complex, global issue that affects every region, every economy, and every future generation.

Addressing it will require sustained effort, accountability, and the political will to match ambition with action.

The time to act was yesterday. The opportunity to change remains today.

Appendix A
UNFCCC Classifications

The United Nations Framework Convention on Climate Change (UNFCCC) divides countries into categories based on their economic development and specific responsibilities.

These classifications are important because they determine what obligations each country has under climate agreements like the Kyoto Protocol and the Paris Agreement.

This appendix explains these groupings and the relevant guiding principles behind them.

UNFCCC Article 2: The Objective of the Convention

The ultimate aim of the UNFCCC, as defined in Article 2, is:

"to stabilize greenhouse gas concentrations in the atmosphere at a level that would prevent dangerous anthropogenic interference with the climate system."

This principle guides how countries are classified and what responsibilities they carry under the Convention and subsequent agreements.

Annex I Parties

Annex I countries are primarily developed nations and economies in transition (former Soviet and Eastern European countries). These countries are expected to take the lead in reducing greenhouse gas emissions and reporting their climate actions regularly.

Examples include:

- United States
- Canada
- European Union countries

- Japan
- Australia
- New Zealand
- Russia
- Ukraine

Annex II Parties

Annex II is a subgroup of Annex I countries. These are developed nations specifically tasked with providing financial and technological support to developing countries to help them mitigate and adapt to climate change.

Examples include:

- United States
- Canada
- European Union countries
- Japan
- Australia
- New Zealand

Note: Economies in transition (such as Russia and Ukraine) are part of Annex I but not Annex II.

Non-Annex I Parties

Non-Annex I countries are mostly developing nations. They have no binding emission reduction targets under the Kyoto Protocol but are encouraged to pursue sustainable development and report periodically on their climate actions. Many of these countries are highly vulnerable to the impacts of climate change.

Examples include:

- India
- China

- Brazil
- South Africa
- Pakistan
- Indonesia
- Most African, Latin American, and Southeast Asian nations

Annex B Parties

Annex B lists the countries under the Kyoto Protocol that committed to specific, legally binding emission reduction targets for the period 2008–2012. The commitments vary by country, based on their historical contribution to greenhouse gas emissions.

Examples include:

- United States (originally signed but later withdrew)
- European Union countries
- Japan
- Canada (later withdrew)
- Russia
- Australia
- New Zealand

Non-Annex B Parties

Non-Annex B countries under the Kyoto Protocol did not have binding emission reduction targets during the 2008–2012 period. These countries were largely developing nations, many of which contributed little to historical emissions but are often the most vulnerable to climate change impacts.

Examples include:

- India
- China
- Brazil

- South Africa
- Most countries in Africa, Latin America, and Southeast Asia

Least Developed Countries (LDCs)

The Least Developed Countries (LDCs) are a group of countries identified by the United Nations as having the lowest indicators of socioeconomic development. LDCs are particularly vulnerable to the impacts of climate change due to limited financial and technical resources. Under the UNFCCC, LDCs receive special support for adaptation and capacity-building.

Examples include:

- Bangladesh
- Ethiopia
- Nepal
- Haiti
- Chad
- Mozambique

APPENDIX B: ACRONYMS/ABBREVIATONS

Acronyms

- **BSTA**: Body for Scientific and Technological Advice
- **GEF**: Global Environment Facility
- **SBI**: Subsidiary Body for Implementation
- **UNFCCC**: United Nations Framework Convention On Climate Change
- **AGCM**: Atmospheric General Circulation Model
- **AIS**: Antarctic Ice Sheet
- **AMO**: Atlantic Multidecadal Oscillation
- **AMOC**: Atlantic Meridional Overturning Circulation
- **AMSU**: Advanced Microwave Sounding Unit
- **AO**: Arctic Oscillation
- **AOD**: Aerosol Optical Depth
- **AR**: Atmospheric River
- **AW**: Atlantic Water
- **BAMS**: Bulletin of American Meteorological Society
- **BC**: Black Carbon
- **BCE**: Before Common Era
- **CAM5**: Community Atmospheric Model Version 5
- **CAPE**: Convective Available Potential Energy
- **CCN**: Cloud Condensation Nuclei

- **CCSM3**: Community Climate System Model Version 3
- **CDR**: Carbon Dioxide Removal
- **CE**: Common Era
- **CENRS**: Committee on Environmental Natural Resources and Sustainability (National Science and Technology Council, White House)
- **CESM-LE**: Community Earth System Model Large Ensemble Project
- **CFCs**: Chlorofluorocarbons
- **CI**: Climate Intervention
- **CMIP5**: Coupled Model Intercomparison Project, Fifth Phase (also CMIP3 and CMIP6)
- **CONUS**: Contiguous United States
- **CP**: Central Pacific
- **CSSR**: Climate Science Special Report
- **DIC**: Dissolved Inorganic Carbon
- **DJF**: December-January-February
- **DOD SERDP**: U.S. Department of Defense, Strategic Environmental Research and Development Program
- **DOE**: U.S. Department of Energy
- **EAIS**: East Antarctic Ice Sheet
- **ECS**: Equilibrium Climate Sensitivity
- **ENSO**: El-Nino-Southern Oscillation
- **EOF Analysis**: Empirical Orthogonal Function Analysis
- **EP**: Eastern Pacific
- **ERF**: Effective Radiative Forcing
- **ESD**: Empirical Statistical Downscaling

- **ESDM**: Empirical Statistical Downscaling Model
- **ESM**: Earth System Model
- **ESS**: Earth System Sensitivity
- **ETC**: Extra Tropical Cyclone
- **ETCCDI**: Expert Team on Climate Change Detection Indices
- **GBI**: Greenland Blocking Index
- **GCIS**: Global Change Information System
- **GCM**: Global Climate Model
- **GeoMIP**: Geoengineering Model Intercomparison Project
- **GFDL HIRAM**: Geophysical Fluid Dynamics Laboratory, global high Resolution Atmospheric Model, (NOAA)
- **GHCN**: Global Historical Climatology Network (National Centers for Environmental Information, NOAA)
- **GHG**: Greenhouse Gas
- **GMSL**: Global Mean Sea Level
- **GMT**: Global Mean Temperature
- **GPS**: Global Positioning System
- **GRACE**: Gravity Recovery And Climate Experiment
- **GrIS**: Greenland Ice Sheet
- **GWP**: Global Warming Potential
- **HadCM3**: Hadley Centre Coupled Model, Version 3
- **HadCRUT4**: Hadley Centre Climatic Research Unit Gridded Surface Temperature Dataset 4
- **HCFCs**: hydrochlorofluorocarbons
- **HFCs**: hydrofluorocarbons
- **HOT**: Hawai'i Ocean Time Series
- **HOT-DOGS**: Hawai'i Ocean Time Series Data Organization &

Graphical System

- **HURDAT2**: revised Atlantic Hurricane Database (National Hurricane Center, NOAA)
- **IAM**: Integrated Assessment Model
- **IAV**: Impacts, Adaptation, and Vulnerability
- **INMCM**: Institute for Numerical Mathematics Climate Model
- **IPCC**: Intergovernmental Panel on Climate Change
- **IPCCAR5**: Fifth Assessment Report of the IPCC; also, SPM – Summary for Policymakers, and WG1, WG2, WG5 – Working Groups 1-3
- **IPO**: Interdecadal Pacific Oscillation
- **IVT**: Integrated Vapor Transport
- **JGOFS**: U.S. Joint Global Ocean Flux Study
- **JJA**: June-July-August
- **JTWC**: Joint Typhoon Warning Centre
- **LCC**: Land Cover Changes
- **LULCC**: Land Use and Land Cover Changes
- **MAM**: March-April-May
- **MSU**: Microwave Sounding Unit
- **NAM**: Northern Annular Mode
- **NAO**: North Atlantic Oscillation
- **NARCCAP**: North American Regional Climate Change Assessment Program (World Meteorological Organization)
- **NAS**: National Academy of Sciences
- **NASA**: National Aeronautics and Space Administration
- **NCA**: National Climate Assessment
- **NCA3**: Third National Climate Assessment

- **NCA4**: Fourth National Climate Assessment
- **NCEI**: National Centers for Environmental Information (NOAA)
- **NDC**: Nationally Determined Contribution
- **NOAA**: National Oceanic and Atmospheric Administration
- **NPI**: North Pacific Index
- **NPO**: North Pacific Oscillation
- **NPP**: Net Primary Production
- **OMZs**: Oxygen Minimum Zones
- **OSTP**: Office of Science and Technology Policy (White House)
- **PCA**: Principal Component Analysis
- **PDO**: Pacific Decadal Oscillation
- **PDSI**: Palmer Drought Severity Index
- **PETM**: Paleo-Eocene Thermal Maximum
- **PFCs**: Perfluorocarbons
- **PGW**: Pseudo Global Warming
- **PNA**: Pacific North American Pattern
- **RCM**: Regional Climate Models
- **RCP**: Representative Concentration Pathway
- **RF**: Radiative Forcing
- **RFaci**: aerosol-cloud interaction (effect on RF)
- **RFari**: aerosol-radiation interaction (effect on RF)
- **RMSE**: Root Mean Square Error
- **RSL**: Relative Sea Level
- **RSS**: Remote Sensing Systems
- **SO6**: surface-to-6 km layer

- **SCE**: Snowy Cover Extent
- **SGCR**: Subcommittee on Global Change Research (National Science and Technology Council, White House)
- **SLCF**: Short Lived Climate Forcer
- **SLCP**: Short Lived Climate Pollutant
- **SLR**: Sea Level Rise.
- **SOC**: Soil Organic Carbon
- **SRES**: IPCC Special Report on Emissions Scenarios
- **SREX**: IPCC Special Report on Managing the Risks of Extreme Events and Disasters to Advance Climate Change Adaptation
- **SRM**: Solar Radiation Management
- **SSC**: Science Steering Committee
- **SSI**: Solar Spectral Irradiance
- **SSP**: Shared Socioeconomic Pathway
- **SST**: Sea Surface Temperature
- **STAR**: Center for Satellite Applications and Research (NOAA)
- **SWCRE**: Shortwave Cloud Radiative Effect (on radiative fluxes)
- **LWCRE**: Longwave Cloud Radiative Effect (on radiative fluxes)
- **TA**: Total Alkalinity
- **TC**: Tropical Cyclone
- **TCR**: Transient Climate Response
- **TCRE**: transient climate response to cumulative carbon emissions
- **TOPEX/JASON1,2**:Topography Experiment/Joint Altimetry Satellite Oceanography Network satellites (NASA)
- **TSI**: Total Solar Irradiance
- **TTT**: Temperature Total Troposphere
- **UAH**: University of Alabama, Huntsville

- **UHI**: Urban Heat Island (effect)
- **UNFCCC**: United Nations Framework Convention on Climate Change
- **USGCRP**: U.S. Global Change Research Program
- **USGS**: U.S. Geological Survey
- **UV**: ultraviolet
- **VOCs**: Volatile Organic Compounds
- **WAIS**: West Antarctic Ice Sheet
- **WCRP**: World Climate Research Programme
- **WMGHG**: Well-Mixed Greenhouse Gas
- **WOCE**: World Ocean Circulation Experiment (JGOFS)
- **EIT**: Economies In Transition
- **OECD**: Organization for Economic Cooperation and Development
- **LDCs**: Least Developed Countries

Abbreviations and Units

- **C**: Carbon
- **CO**: carbon monoxide
- **CH4**: methane
- **cm**: centimeters
- **CO_2**: carbon dioxide
- **°C**: degrees Celsius
- **°F**: degrees Fahrenheit
- **GtC**: gigatonnes of carbon
- **hPA**: hectopascal
- **H_2S**: hydrogen sulfide

- **H_2SO_4**: sulfuric acid
- **Km**: kilometers
- **m**: meters
- **mm**: millimeters
- **Mt**: megaton
- **µatm**: microatmosphere
- **N**: nitrogen
- **N_2O**: nitrous oxide....
- **NOx**: nitrogen oxides
- **O_2**: molecular oxygen
- **O_3**: ozone
- **OH**: hydroxyl radical
- **PgC**: carbon petagrams
- **ppb**: parts per billion
- **ppm**: parts per million
- **SF_6**: sulfur hexafluoride
- **SO_2**: sulfur dioxide
- **TgC**: teragrams of carbon
- **W/m^2**: Watts per meter squared

APPENDIX C
USEFUL WEBSITES

United Nations Framework Convention on Climate Change.

https://unfccc.int

Intergovernmental Panel on Climate Change (IPCC)

http://www.ipcc.ch

United Nations Climate Action

https://www.un.org

National Centers for Environmental Information (NOAA)

www.ncei.noaa.gov

Center for Climate and Energy Solutions

https://www.c2es.org

National Center for Atmospheric Research (National Science Foundation)

https://ncar.ucar.edu

Global Change (U.S. Global Change Research Program)

http://www.globalchange.gov

Made in the USA
Monee, IL
03 November 2026

67386707R00066